SpringerWienNewYork

CISM COURSES AND LECTURES

The series presents lecture notes, monographs, edited works and proceedings in the field of Mechanics, Engineering, Computer Science and Applied Mathematics.
Purpose of the series is to make known in the international scientific and technical community results obtained in some of the activities organized by CISM, the International Centre for Mechanical Sciences.

INTERNATIONAL CENTRE FOR MECHANICAL SCIENCES

COURSES AND LECTURES - No. 514

COMPUTATIONAL AND EXPERIMENTAL MECHANICS OF ADVANCED MATERIALS

EDITED BY

VADIM V. SILBERSCHMIDT
LOUGHBOROUGH UNIVERSITY, LOUGHBOROUGH, LEICESTERSHIRE, GREAT BRITAIN

SpringerWienNewYork

This volume contains 106 illustrations

SPIN 12753598

All contributions have been typeset by the authors.

ISBN 978-3-211-99684-3 SpringerWienNewYork

PREF CE

Advanced materials (composites, multiphase materials, materials for microelectronics, biomaterials, etc.) play a crucial role in modern engineering and biomechanical applications where they are often exposed to complex loading and environmental conditions. In many cases, new approaches are needed to characterise various features of these materials and to model their deformational behaviour, failure processes as well as to analyse reliability of components and structures under different conditions. Such approaches should be calibrated and validated by specific experimental techniques, quantifying both microstructural features and respective mechanisms at various length scales. The aim of the course is to give an overview of various modelling tools and experimental methods that can be employed to analyse and estimate properties and performance of advanced materials.

The first paper (by P.D. Ruiz and V.V. Silberschmidt) deals with experimental analysis of mechanical behaviours of advanced materials. A special emphasis is on techniques used to gain quantitative microstructural information needed for material modelling and/or experimental validation of model predictions. Two main groups of techniques are covered - mechanical tests and full-field analysis. The paper starts with application of dynamic mechanical analysis to viscoelastic materials followed by a discussion of experimentation with microspecimens. Theoretical aspects of nanoindentation are presented together with case studies for a Ni-based alloy and ceramic coating. A review of full-field measuring techniques is accompanied by examples of their use to validate predictions of stresses in adhesive joints and surface strains, damage detection and characterisation as well as identification of material's parameters.

Another part of the course deals with modern theoretical approaches used to analyse heterogeneous materials and a non-linear material behaviour. The first topic is covered in a paper by G.S. Mishuris, A.B. Movchan and L.I. Slepyan that reviews main results obtained for periodic structures based on dynamic lattice Green's functions. Among the analysed phenomena are localization of vibrations near defects and fracture in structured media, including wave tunnelling along the crack and energy dissipation in the lattice containing a moving crack. In the next paper (by M. Jabareen and M.B. Rubin) the second topic is studied by means of the formulation of a 3-D brick Cosserat Point Element (CPE) for the solution of problems in non-linear elasticity. The paper opens with a review of some tensor operations and kinematic measures in continuum mechanics followed by introduction of the CPE that exhibits no locking for nearly incompressible materials and for thin structures, like plates and shells. Some example problems for such structures are discussed.

A paper by E. Busso reviews multiscale materials modelling approaches required to complement continuum and atomistic analyses methods. Their potential for computational materials design, based on the understanding of the dual nature of the structure of matter - continuous when viewed at large length scales and discrete when viewed at an atomic scale - is discussed. Various methods of continuum mechanics are reviewed, including a local crystallographic framework and non-local approaches. The paper finishes with a discussion of the problem of bridging the length and time scales. Another paper (by H.J. Böhm, D.H. Pahr and T. Daxner) deals with numerically based continuum modelling of thermomechanical and thermophysical behaviours of microstructured materials. After a review of mean-field methods and variational bounds the authors discuss various aspects of, and methods for, modelling discrete microstructures. Among the presented applications are elastoplastic composites at finite strains, diamond particle-reinforced composites as well as porous and cellular materials.

In the final paper by V.V. Silberschmidt, several case studies are analysed in order to demonstrate the strategies used to solve the real-life problems, in which the microstructure of materials directly a ects their response to loading and/or environmental conditions. Among the presented examples are studies of the e ect of microstructural randomness of carbon fibre-reinforced composites and ceramic coatings on their properties and performance as well as incorporation of microstructure and various deformational mechanisms in models of flip chip microelectronic packages.

The book is addressed to doctoral students, young researchers as well as practicing R&D engineers, dealing with advanced materials, components and structures.

Vadim V. Silberschmidt

ONTENTS

Experimental Analysis of Mechanical Behaviour of Advanced Materials

Pablo D. Ruiz and Vadim V. Silberschmidt

Wolfson School of Mechanical and Manufacturing Engineering, Loughborough University, UK

1 Introduction

A notion of a material's mechanical property in many cases is not an unambiguous one. There are two main groups of reasons for this. The first is linked to *conditions* of the test, the second to the *length scale*, for which property is sought. Hence, instead of a narrow understanding of a mechanical property as a constant magnitude obtained from the handbook or a database, modern researchers dealing with real-life applications of advanced materials should rather consider it as a multi-parametric function and, in some cases, even a statistic one.

Let us start with analysis of the first group since it has direct implications for performing the tests to obtain material's properties. The effect of *environmental* conditions on mechanical parameters of materials is mostly appreciated in terms of temperature-dependent properties. Other environmental factors could be also important for some specific applications (e.g. hygroscopic effect in composites, exposure to aggressive environment, irradiation effects etc.) but they are outside the scope of this Chapter. Another important factor is the *loading rate*, affecting non-elastic response of various materials, e.g. strain-rate hardening in plasticity and strain-rate sensitivity of a viscous behaviour. A *type of loading* is also important, especially for strength assessment – consider the differences between the (quasi)static strength, on the one hand, and dynamic (impact) and cyclic loading (fatigue) strengths, on the other. A *loading state*, used in experiments, is usually supposed to be simple and permanent (i.e. non-changing during the experiment); an implementation of these requirements can become rather cumbersome in cases of large deformations.

An importance of the *length scale* is linked to microscopic heterogeneity of most materials. This can be complicated by a presence of several levels of nonuniformity, e.g. in layered structures with different properties of layers, functionally-graded materials etc. Hence, two levels of properties assessment are possible. For specimens larger than the size of *representative volume element* (*RVE*), or, in other words, containing a large representative set of microstructural features (grains, microdefects, reinforcing elements in composites etc.) – *effective* (sometimes called *global* or *homogenized*) properties are obtained in tests. These magnitudes are usually presented in handbooks, databases etc. of material's properties. In contrast, properties measured at the lower length scale – known as *local properties* – demonstrate a considerable spatial scatter. The extent of the latter depends on the scale size (*window size*) – the smaller the window, the larger the scatter, which, in the case of two-phase materials will have two natural bounds linked to the properties of the constituents (Silberschmidt, 2008).

In many practical cases, the global properties cannot be used in models due to a large size of the RVE compared to characteristic dimensions of a component. Besides, a transition to critical and post-critical behaviour of materials that is characterised by the onset of plastic flow, macroscopic damage and/or failure is usually linked to *localisation* of these mechanisms precluding the use of effective parameters. One of the obvious examples is a transition from a macroscopically uniform deformation of a large specimen of polycrystalline material in the elastic regime to a spatially localised plastic flow, e.g. due shear band formation starting in grains that are appropriately oriented with respect to maximum shear stresses.

All the discussed features should be adequately accounted for in preparation of test programs and employment of specific techniques. This Chapter can by no means present a fully comprehensive description of all experimental techniques used to assess mechanical behaviour of materials. It rather deals with some specific types of tests that are usually not covered by standard textbooks on experimental mechanics dealing mainly with global properties. Since descriptions of the latter are (to some extent) ubiquitous, the authors decided to narrow down the choice of the techniques to those that either characterise more advanced properties (e.g. relaxation ones), local (microstructure-induced) properties or their spatial distribution – so called *full-field* methods.

2 DMA & visco-elastic properties

In contrast to elastic and plastic behaviour, characterisation of the visco-elastic one is still not a fully established in the engineering community. This can be explained by historic reasons linked to a relatively late broad introduction of plastics, demonstrating such behaviour, into various products, as compared to traditional structural materials. High-temperature applications with a long-term exposure of components (made predominantly, of metals and alloys) to extreme thermal conditions became ubiquitous nearly at the same time. Another reason is unsuitability of standard tensile tests to provide a direct information about parameters characterising viscous behaviour in the same (relatively) simple way as for the elastic behaviour (Young's modulus) or the plastic one (yield stress, hardening modulus). Special tests – creep (loading with a constant stress level) and relaxation (loading with a constant deformation) – that can be nowadays performed with the same universal testing machines used for tensile testing are rather time-consuming when a full description of visco-elasticity is necessary.

Various applications can not be reduced to the cases of pure creep or relaxation; among the obvious examples are solder points/balls in microelectronic packages exposed to creep-fatigue under service conditions or plastics and composites under dynamic (e.g. impact) loading. The need for the parameters characterising the visco-elastic behaviour is additionally enhanced by expansion of mechanics of materials into domain of biomaterials, many of which can be considered as visco-elastic media.

A special type of tests – *dynamic mechanical analysis* (DMA) – was developed to characterise visco-elastic properties of materials. Let us start with some general ideas of visco-elasticity that will be analysed following classical books (Christensen, 1971 and Tschoegl, 1989).

The visco-elastic behaviour is often described in terms of a so-called *hereditary* approach with stress being dependent on the current strain state as well as the history of strain in the material. For a uniaxial loading at $t = 0$ of an isotropic linear visco-elastic material under isothermal conditions the constitutive equation can be introduced in terms of a hereditary integral (also known as *Boltzmann superposition* integral):

$$\sigma(t) = \int_{0_-}^{t} E(t-\tau)\dot{\varepsilon}(\tau)d\tau , \tag{1}$$

where $E(t)$ is a relaxation function, $\dot{\varepsilon}$ is a strain rate. Another form can be used to explicitly introduce the strain history:

$$\sigma(t) = E_e\left(\varepsilon(t) - \int_{0_-}^{t} \hat{E}(t-\tau)\dot{\varepsilon}(\tau)d\tau\right), \tag{2}$$

where E_e is an equilibrium modulus.

For the so called *generalised Maxwell model* (also known as *Wiechert model*) consisting of a spring and N Maxwell elements (see Figure 1) the relaxation

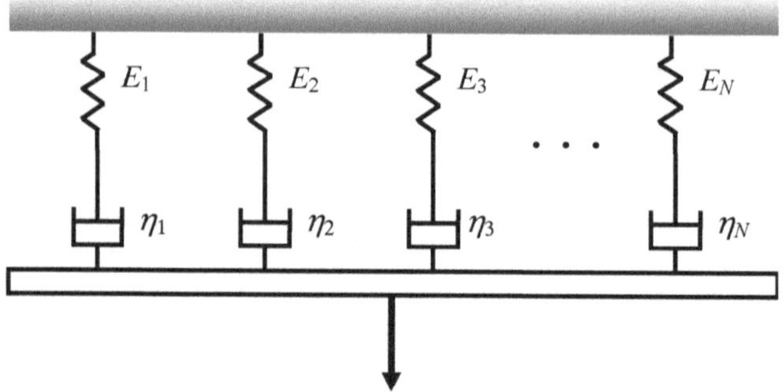

Figure 1. Generalised Maxwell model.

model has the following form:

$$E(t) = E_e + \sum_{i=1}^{N} E_i e^{-\frac{t}{t_{ti}}}, \tag{3}$$

where $t_i = \eta_i / E_i$ are relaxation times of elements. The second term is also known as the *Prony series*.

Using the Laplace transformation

$$\bar{f}(s) = \int_0^\infty f(t) e^{-st} dt, \tag{4}$$

an *operational modulus* (Park and Schapery, 1999) can be obtained in the following form:

$$\overline{E}(s) = s \int_0^\infty E(t) e^{-st} dt. \tag{5}$$

The frequency-dependent version of the stress relaxation modulus can be obtained for a steady-state harmonic (sinusoidal) loading with a frequency of oscillation ω:

$$E^*(\omega) = \overline{E}(s)\Big|_{s \to i\omega} = E'(\omega) + iE''(\omega), \tag{6}$$

where E' and E'' are the so called storage and loss moduli.

A more clear analysis can be obtained by comparison of the responses of different types of media to the harmonic excitation presented in Figure 2. Obviously, in a perfectly elastic media (Figure 2a) phases of strain and stress diagrams would coincide, since both strain and stress are directly proportional due to the Hooke's law. A perfectly viscous material is characterised by the known relationship between the stress and the strain rate: $\sigma = \eta \dot{\varepsilon}$ that means that in a system with harmonic strain excitation the stress level will change with a delay by $\varphi = \pi/2$ (Figure 2b).

A linear visco-elastic material is an intermediate case: for a harmonic excitation with strain the stress changes will be characterised by the phase difference δ (Figure 2c) that due to standard dynamic analysis (Inman, 2007)

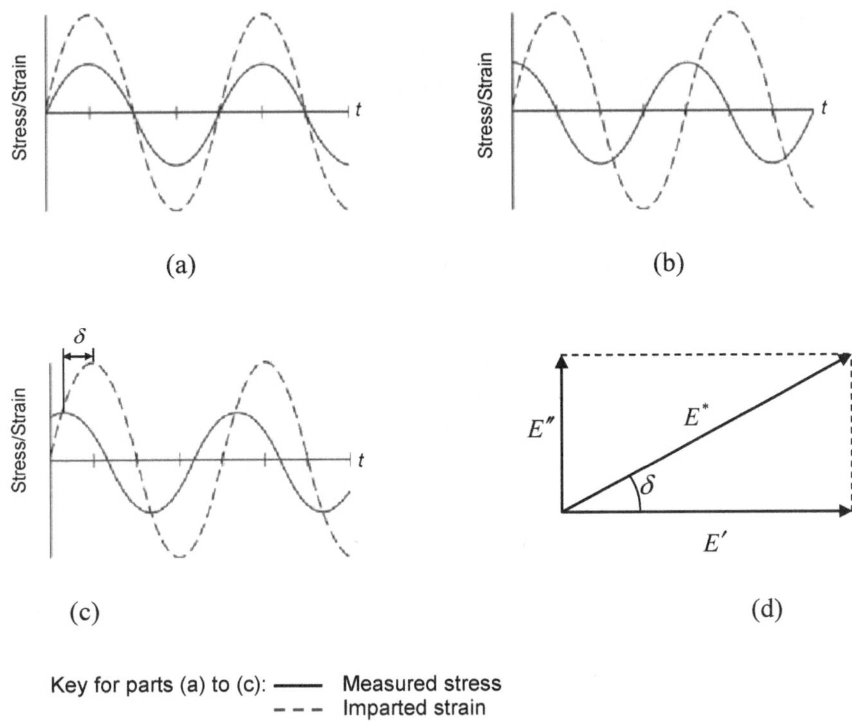

Figure 2. Cyclic stress strain diagrams for (a) elastic, (b) viscous and (c) visco-elastic materials, (d) complex modulus diagram for visco-elastic material.

can be determined as

$$\delta = \tan^{-1}\frac{E''}{E'}, \tag{7}$$

(see the respective *Argand diagram* in Figure 2d). The phase difference δ indicates the extent of material's viscosity, with $\delta = 0$ and $\delta = \pi/2$ being equivalent to perfectly elastic and perfectly viscous materials, respectively.

Generally, a storage modulus E' specifies the energy stored in the specimen due to the applied strain, and a loss modulus E'' specifies the dissipation of energy as heat. A useful quantity is the *damping factor* or *loss tangent*

$$\tan\delta = \frac{E''}{E'}, \tag{8}$$

representing a part of mechanical energy dissipated as heat during the loading/unloading cycle. Apparently, it vanishes for a perfectly elastic material and is equal to infinity for a perfectly viscous one. This parameter is linked to the so called *quality factor*, or *Q factor*:

$$Q = \frac{1}{\tan\delta}. \tag{9}$$

The above discussion deals with fundamentals of DMA. In dynamic mechanical analysis tests a harmonic excitation is used with a simultaneous measurement of the amplitudes of stresses and strains as well as the phase angle δ between them. This excitation is low-amplitude so that no plastic deformation could affect the results. Various loading schemes can be used – tensional, torsional and shear. Most materials demonstrate dependence of E', E'' and $\tan\delta$ on frequency and temperature. So, modern DMA testing machines usually perform a test with automatically changing frequencies and temperatures. The latter is important, for instance, for estimation of additional properties such as the glass transitions in polymers.

The scheme overcomes limitations of transferability of results based on the creep and relaxation approach to short-term and high-frequency ranges that are characteristic for various service conditions (Christensen, 1971). As an example, Figure 3 presents results of DMA testing of constituent layers of an adidas football (for details see (Price et al., 2008)). This type of data obtained with DMA can be used to model the visco-elastic behaviour of materials for arbitrary loading conditions and histories. The finite-element software package ABAQUS, for instance, transfers the data for storage and loss moduli into the Prony series (Abaqus, 2003). The latter in the frequency domain will have the following form

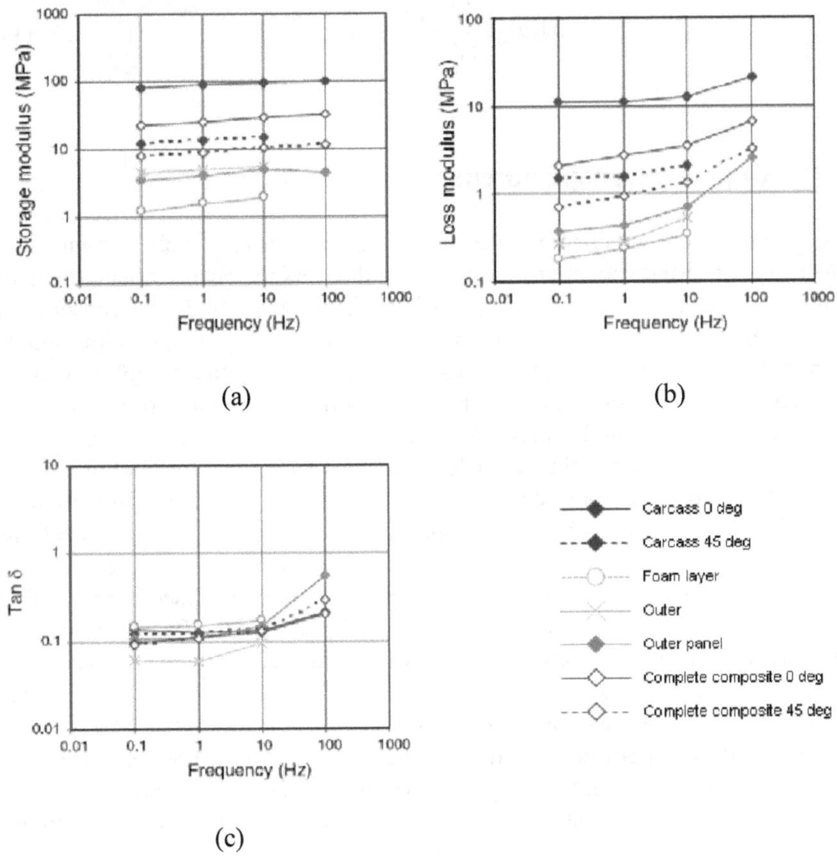

Figure 3. Visco-elastic properties of constituent layers of adidas ball: (a) Storage modulus, (b) loss modulus and (c) $\tan\delta$.

(Park and Schapery, 1999):

$$E'(\omega) = E_e + \sum_{i=1}^{N} \frac{\omega^2 t_{ri}^2}{1+\omega^2 t_{ri}^2} E_i \tag{10}$$

and

$$E''(\omega) = \sum_{i=1}^{N} \frac{\omega t_{ri}}{1+\omega^2 t_{ri}^2} E_i \,. \tag{11}$$

3 Testing of microspecimens

A scale-dependent mechanical behaviour of materials is well known. With dimensions of tested specimens approaching those of the characteristic elements of microstructure, an averaging effect, which 'smears' both the presence of microstructural features as well as realisation of various deformational mechanisms at lower length scales over a large (macroscopic) volume, diminishes. This can result in significant deviations of the local properties from the global (i.e. averaged) ones. A detailed study of the local properties is important in many cases linked to localisation of deformation and/or fracture processes. Another obvious reason for such studies is linked to applications of components/structures with microscopic dimensions, e.g. in microelectronic packages. In this case, direct testing of microspecimens is unavoidable.

A typical study of lead-free solders at micro scale is analysed here based on works by Gong et al. (2006a, 2007a) and Gong (2007). A process of miniaturisation of microelectronic devices leads to a continuous decrease in the dimensions of all microelectronic components. An additional need to increase the number of interconnections between elements of packages has been driven a diminishment of solder bumps to dimensions below 100 μm. Elements with such dimensions can contain only few grains, making the use of the data obtained for bulk macroscopic polycrystalline specimens at best questionable. To overcome this hindrance, a program of testing at a micro scale was undertaken. A material of interest – lead-free SnAgCu solder – was used in specially prepared specimens. The latter consisted of two Cu plates (15 mm × 15 mm, thickness 1 mm) soldered together by a thin layer of the studied material (Figure 4). A special device was used to change the thickness of the solder layer; in the tests two thicknesses – 100 μm and 1000 μm – were used. The reflow process that transformed the solder paste into solders was implemented with a Planer T-TRACK© reflow oven; two different cooling rates were used to study their effect on the microstructure and properties of solder joints.

After the reflow the specimen was cut with a low-speed diamond saw into specimens with the cross-section 1 mm × 1 mm (see Figure 4b) that were used in mechanical tests. Such specimens can be directly tested in tension using Instron MicroTester. In microelectronic packages solder bumps are exposed to temperature changes resulting in shear deformations due to the mismatch in coefficients of thermal expansion of constituent materials. So, a special setup

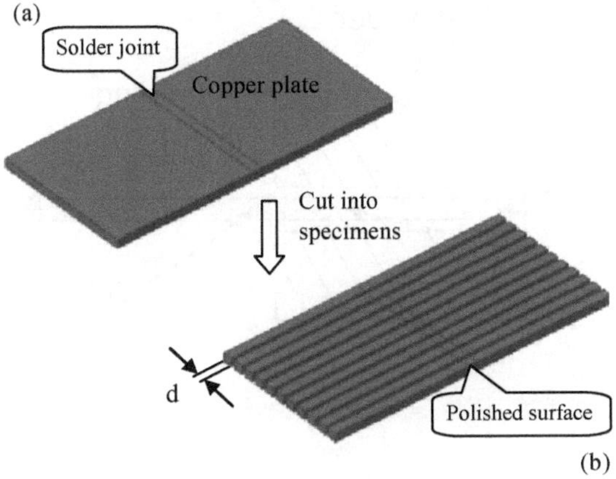

Figure 4. Preparation of microspecimens: (a) specimen after soldering; (b) specimens for the mechanical test. d ≈ 1 mm.

was prepared for testing the manufactured specimens (Figure 5).

A steel holder with a hole, the cross section of which is slightly larger than that of the specimen, is used. It is fixed on one of the testing machine's grips. One of Cu substrates of the specimen is inserted into the hole as shown in Figure 5 while the other substrate at the opposite end is fixed to the machine's second grip, the movement of which is controlled by the MicroTester. Then the hole in the steel holder is filled with Epoxy resin, making sure that the resin does not contact with the solder joint. After 24 hours of curing, the epoxy resin is fully hardened, and the specimen is assembled in the MicroTester in a stress-free state.

In the shear test, the solder joint yields with the copper substrates being still in the elastic state since its yield point is much lower than that of pure copper. At the same time, the Young's modulus of Cu is significantly higher and effectively all the deformation is localised in the solder material. Therefore, the applied engineering shear strain rate $\dot{\gamma}$ on the joint is approximately:

$$\dot{\gamma} = \frac{\dot{U}}{b}. \tag{12}$$

where $\dot{U}$ is the applied displacement rate of the grip and b is the distance

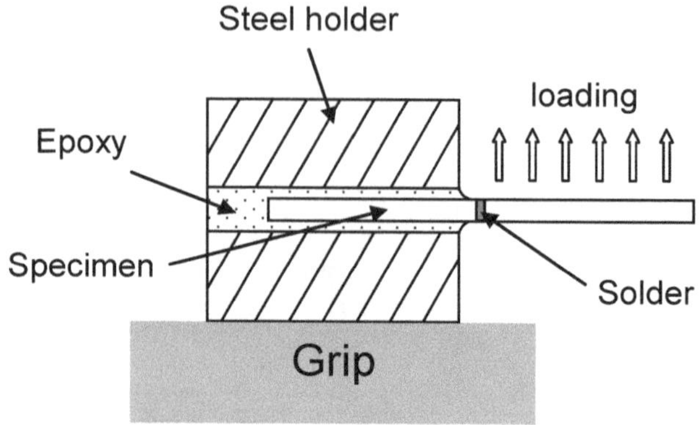

Figure 5. Setup for shear test.

between substrates (it is the distance between the middle points of interfaces for the real specimens). A resin with the highest available stiffness was used in order to reduce its effect on the results.

The mechanical test was performed with the loading rate 0.1 μm/s that corresponds to the shearing strain rate is 1×10^{-3} s^{-1} for solder thickness 100 μm. The specimen at first is loaded for a specific time to reach a prescribed shearing deformation then held in the tester for 24 hours to release the residual stresses.

The microstructure of solder was studied before and after the deformation; both scanning electron microscopy (SEM) and an optical microscopy in polarized light (PL) were employed to examine grain features. Additionally, transmission electron microscopy (TEM) was used to study crystallographic parameters of local areas after loading and to characterize the deformation and damage behaviour of substructures within SnAgCu grains.

A typical result of shear deformation in the solder specimen is presented in Figure 6. Apparently, only two grains occupy the entire cross section of the solder. Their mechanical behaviour is determined by their crystallographic orientation: slip bands inside each grain have the same direction; however, there is a large angle between these bands for the two grains. Shear of stiff, parallel Cu plates induces practically the (macroscopically) uniform deformation state in the solder (with some deviations near the specimen's edges and grain boundaries) of these two grains. Hence, different orientations of shear bands demonstrate the lattice-dependent behaviour and, consequently, anisotropic properties of a SnAgCu single grain. Since a joint in electronic packages can contain one or a

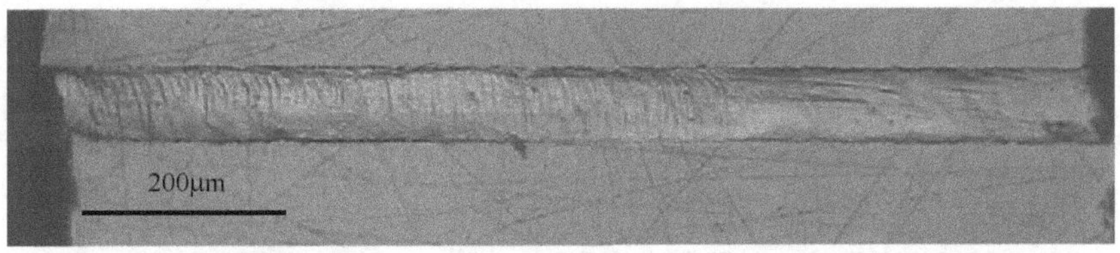

Figure 6. Bright-field image of a loaded joint (b = 82 μm) after shear $\gamma = 39\%$.

few grains, this grain-based behaviour would define its response to the in-service conditions.

The obtained results were used in two-scale finite element analysis of thermal cycling of lead-free joints (Gong et al., 2006b, 2007b and 2008); these results are presented in another Chapter.

In many cases there is a need to analyse even smaller specimens. For instance, the above solder in the test described above still contains more than one grain. To study the intra-grain deformation another technique was used (for details of the method below see (Gong, 2007)). It employed the *focussed ion beam* (FIB) to cur a specimen from the single grain of the solder as shown in Figure 7a. FIB was used to mill the specimen with length and thickness of 25 μm and 2 μm on the ground and polished solder block obtained by the reflow process with a relatively low cooling rate. The bottom of the micro-specimen was separated from the bloc in order to reduce the effect of the bulk specimen and induce a pure tensile deformation state.

This design allows overcoming another limitation of the previous technique – a (relatively) high loading rate. The latter is defined by the minimum displacement rate of the testing machine (that is 0.1 μm/s) and the size of the solder specimen. Since the entire grips' displacement was applied to the solder, in case of the solder layer thickness 100 μm the lowest possible strain rate was 10^{-3} s^{-1}; for the length of the studied micro-specimen it would be even higher – 4×10^{-3} s^{-1}. The use of the bulk solder specimens incorporating the milled micro-specimen can significantly overcome this limit by increasing the length of the bulk specimen (it was 25 mm in the study). With a (nearly) uniform axial distribution of strain in the tensile test the extension rate of the micro-specimen is the same as of the entire specimen. Hence, the strain rate in this test is decreased by a factor equal to the ratio of the length of the macro-specimen to that of the micro-specimen (up to 500 in this study).

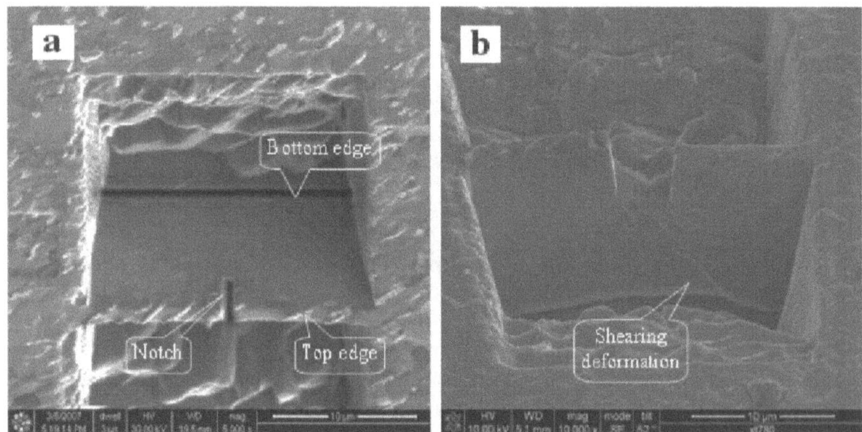

Figure 7. Micro-specimen before (a) and after (b) deformation.

The notch is also introduced into the micro-specimen to study the failure process (Figure 7a). The applied deformation resulted in the strain rate 1×10^{-5} s^{-1}; the specimen after 20% elongation is shown in Figure 7b. The deformed specimen vividly demonstrates the ductile character of deformation – there is no crack growth from the tip of the notch; instead, the notch widens and shear deformation started from the notch at approx. 45° to the tension direction. A further analysis with TEM (gong, 2007) demonstrated that the highly non-uniform character of the deformation process in the single-grain micro-specimen caused by the presence of notch did not only affect the deformation localisation in the form of shear bands but also caused microstructural changes – formation of a low-angle sub-grain boundary.

The two studied cases by no means exhaust possible ways to analyse deformational and failure behaviours at the microscopic scale. Still, they clearly show the considerable increase in the detailedness of information about these behaviours. Besides, a direct effect of microstructure on the spatial realisation of deformational processes and an opposite effect – of the deformation on the microstructure – demonstrate the need for a direct introduction of these processes into respective micromechanical models.

4 Nanoindentation

4.1 Introduction

An apparent difference between the global (macroscopic) and local (microscopic) properties of many advanced materials due to the effects of their microstructure presupposes a development of special techniques for studies at the micro scale. One of the obvious challenges for extension of the conventional testing techniques utilised in universal testing machines to a micro scale is impossibility to perform *in situ* tests in the area of interest. The chance to prepare a micro specimen of the material from this area – though currently feasible with FIB (see above) – still does not fully solve the discussed problem 'specimen *vs.* material' not to mention the complexity of treating such specimens.

Hence, a relatively old idea of the use of indention to measure hardness as a parameter linked to the ratio of the applied force to the effective cross-sectional area of the imprint left by an indentor have become a basis of a new testing technique known as *nanoindentation*. Here, the scaling down is implemented by means of transition to low loads – down to micro-Newtons – resulting in penetration lengths down to nanometres (this is the reason for the name).

Two important moments should be considered here, before a discussion of nanoindentation with its advantages, features and limitations. The first is the use of the free surface of a specimen to measure hardness, and the second is that hardness is *not* a mechanical parameter that can be directly used in any models of mechanics of materials. The former moment presupposes a special attention to preparation of specimens; the latter is alleviated by the use of theories linking hardness to the Young's modulus.

Now, about benefits:

1) Nanoindentation provides a possibility for *en situ* measurements for very small volumes of materials and spatial scale and, hence, can be considered as a *non-destructive* technique.
2) There are options of spatial *scanning* of properties using arrays of indentations points of various configuration and even *continuous* measurements based on nanoscratching (Kaupp and Naimi-Jamal, 2004).
3) A precise placement of the indenting point allows estimation of properties exactly in the area of interest, e.g. in a specific phase, constituent or near any microstructural feature.
4) Automatization of tests allows performing a large testing program, e.g. indentation of the same specimen in many points, without interventions by an operator.

These generic benefits of the method are complementary to the principal feature of nanoindentation, i.e. assessment of properties of materials that otherwise are hardly feasible. Best examples are soft materials (including biological tissues) and thin (and soft) coatings.

Some limitations of the technique will be discussed after the overview of its main principles, which is based on well-known literature (Fischer-Cripps, 2004, Oliver and Pharr, 1992, 2004, and Pharr and Bolshakov, 2002).

4.2 Principles of nanoindentation

Though the history of hardness tests is several centuries long, their use to extract traditional mechanical properties of materials is significantly shorter. This was linked to a transition from the measurement of the residual area of the imprint to a continuous depth-sensing technique in 1960s (see some historical comments in (Borodich and Keer, 2004); it was known under the name *microindentation* at that time).

The depth-sensing method is based on the continuous measurement of the depth of penetration of indentor h and the corresponding level of the applied force P (see Figure 8). In the process of penetration by indentor into a material the maximum attained level of force applied to the indentor P_{max} corresponds to the maximum penetration depth h_{max}. Figure 8 also presents other main parameters of the indent's geometry for an axisymmetric (conical) indentor with a half-angle ϕ: the depth of contact h_c and the corresponding radius of the contact circle a; the final depth h_f; the residual imprint depth after unloading (removing of the indentor); the sink-in dept h_s.

Various types of indentors are used for nanoindentation; one of broadly used ones is the Berkovich indentor – a triangular pyramid, made of diamond. It has

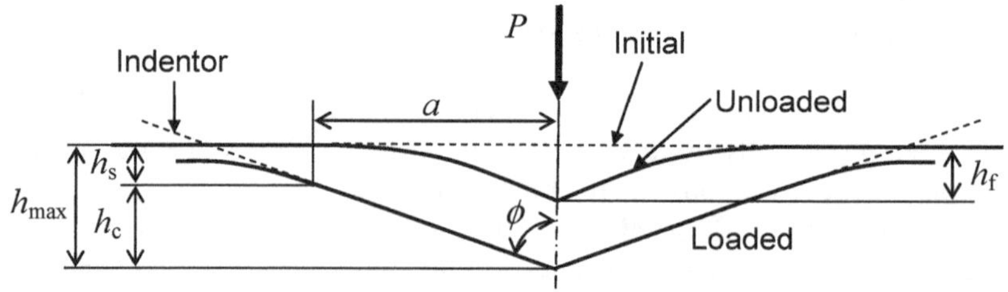

Figure 8. Schematic of indentation process with parameters of indent's geometry (after Oliver and Pharr, 1992).

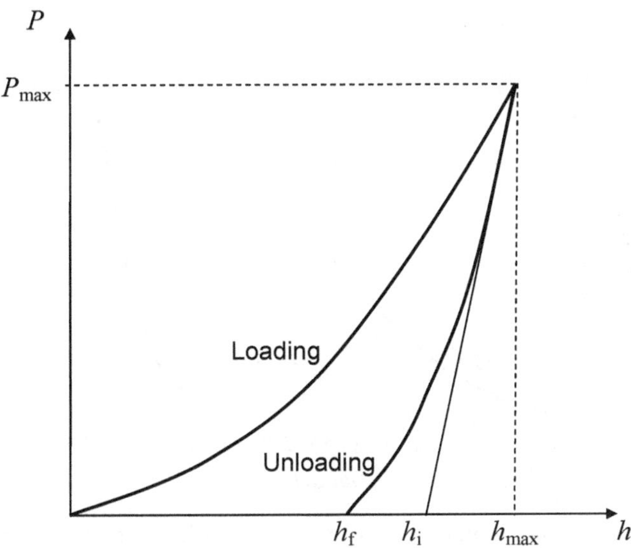

Figure 9. Schematic loading-unloading curve for indentation.

the same area-to-depth relationship as the Vickers indentor, a four-sided pyramid that was used in microindentation tests (Pharr and Bolshakov, 2002). Axisymmetric indentors, including spherical ones, are also used to test various materials.

The continuous measurement of the indentation parameters during the entire cycle of imprinting and removal of the indentor results in the load-displacement (i.e. load-depth) diagram that is schematically presented in Figure 9 (a real curve for a cortical bone tissue is given in Figure 10 (Alam, 2008)). The non-linear loading part of the diagram is followed by an unloading part, situated below the former due to the residual strains caused by the irreversible (plastic in the case of elasto-plastic material) deformation.

The relationships linking the Young's modulus of the tested material and a slope of the initial portion $S = dP/dh$ of the unloading curve were suggested in 1970s (see, e.g. (Bulychev at al., 1975, 1976)). In a more common notation (Oliver and Pharr, 1992, 2004) it can be presented in following form:

$$E_{\text{eff}} = \frac{1}{2\beta}\sqrt{\frac{\pi}{A}}S \tag{13}$$

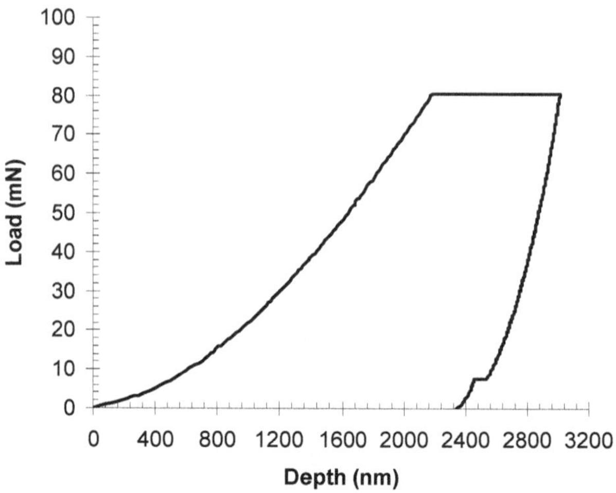

Figure 10. Loading-unloading curve for nanoindentation of cortical bone tissue.

and

$$E_{\text{eff}}^{-1} = \left(1 - \nu^2\right)E^{-1} + \left(1 - \nu_{\text{i}}^2\right)E_{\text{i}}^{-1} . \tag{14}$$

The system of equations (13), (14) can be resolved with regard to the Young's modulus of the tested material E for a known properties of the indentor – its Young's modulus E_{i} and Poisson's ratio ν_{i} – and the given magnitude of the material's Poisson's ratio ν. Other parameters of the system are the cross-sectional area of the indentor A and the dimensionless parameter β. Theoretically, in approximation of infinitesimal strains, $\beta = 1$ for a rigid axisymmetric indentor penetrating an elastic body. The exact magnitude of β is still a highly discussed area; it is known to depend on the shape of indentor, Poisson's ratio, plastic deformation etc. – a discussion on the matter is given in (Oliver and Pharr, 2004) where the magnitude $\beta = 1.05$ was suggested 'as good a choice as any".

The notion of contact area A is very important to assessing the *hardness* (term *nanohardness* is used for nanoindentation results) of the tested material:

$$H = \frac{P_{\max}}{A} . \tag{15}$$

It is important to note that A is *not* an area of the residual imprint; it is linked to the contact area *under load.* In effect, it corresponds to the projection of the area of contact when indentor reaches depth h_{max}, i.e. for a case when the depth, along which there is a contact with the indentor, is equal to h_c (see Figure 8).

The current research into applications of nanoindentation is aimed at assessment of additional mechanical properties or parameters. The level of residual stress σ_R can be assessed using the following equation for a spherical indentor with radius R (Fischer-Cripps, 2004):

$$\sigma_R = \sigma_Y - \left[\frac{4E_{eff}}{3.3\pi}\right]\frac{a}{R}. \tag{16}$$

Nanoindentation can also be used to characterise fracture toughness of materials (Kese and Rowcliffe, 2003), based on the scheme developed for microindentation in 1980s (Lawn et al., 1980). For the known Young's modulus and hardness, the fracture toughness of the tested material can be expressed in the following way:

$$K_c = \alpha\sqrt{\frac{E}{H}}\frac{P}{c_0^{3/2}}, \tag{17}$$

where c_0 is the length of the radial crack, caused by indentation with load P. Parameter α is linked to geometry of indentor (for a cube corner $\alpha = 0.04$ (Field and Swan, 1995)).

The underlying formulae vividly demonstrate limitations of the used approach: this scheme is applicable to isotropic materials with time-independent behaviour. Many novel schemes of applications link results of nanoindentation with detailed finite-element simulations that allow researchers – in many cases by means of inverse analysis – to deal with more complex cases, such as anisotropic materials (Bocciarelli et al., 2005). There are also modifications aimed at alleviation of these limitations – an account for creep is introduced in (Lucas and Oliver, 1999).

4.3 Examples of implementation of nanoindentation

There are various ways to use nanoindentation but – regarding limitations of the method – the most suitable way is to use it for a comparative analysis. The latter can be applied to quantify either the difference in properties of various specimens or the scatter in the spatial distribution of the properties of a single

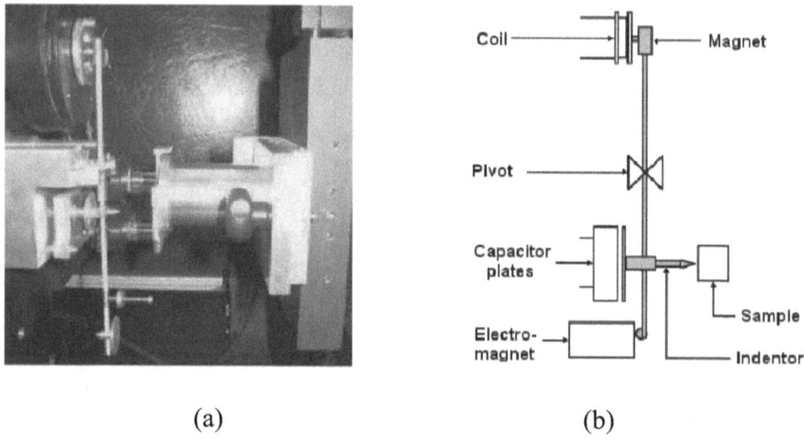

Figure 11. NanoTest Platform: (a) general view; (b) schematic.

material due to its microstructure. This Section presents examples of both approaches.

Effect of Ultrasonically-Assisted Machining. Ultrasonically-Assisted Machining (UAM) is a new technology based on superposition of ultrasonic vibration with amplitude 15-30 μm and frequency around 20 kHz on the movement of the cutting tool (Babitsky et al., 2004). This vibration significantly improves conventional cutting technologies resulting in reduction of cutting forces, improved surface finish etc., allowing turning of hard-to-turn materials such as nickel-based superalloy INCONEL 718. To study the difference between the conventional turning (CT) and ultrasonically-assisted turning (UAT), it is necessary to understand the extent of the effect of technology on the sub-surface layers of machined specimens. Hence, nanoindentation tests were performed with the NanoTest Platform made by Micro Materials Ltd, Wrexham, UK (Figure 11) to estimate the layer affected by turning.

The specification of the range and sensitivity for the displacement and load is shown in Table 1.The Berkovich indenter is used in all the tests.

In NanoTest Platform a load to the indentor is applied by means of a coil and magnet located at the top end of the pendulum. The penetration of the probe into the sample is monitored with a sensitive capacitive transducer. In order to prepare a specimen for the nanoindentation analysis, parts of work pieces that were machined with CT and UAT were placed facing each other and potted into

Table 1. Specification of the NanoTest Platform.

Displacement range and sensitivity	
Range	0-50 μm
Noise-floor	0.025% of full-scale deflection
Theoretical resolution	0.04 nm
Load range and sensitivity	
Maximum resolution	better than 100 nN
Load ranges	up to 0-500 mN
X/Y/Z resolution/travel	0.02 mm/50 mm
Analysis area	50 mm × 50 mm

epoxy resin as it is shown in Figure 12a. After polishing the produced specimen was mounted into the NanoTest machine (Mitrofanov, 2004, and Ahmed et al., 2006). The tests were carried out in an automatic mode with a total duration of two days. There are three runs A, B and C, parallel to each other (vertical lines in Figure 12b). Each of these lines goes from the conventionally machined part of the specimen, crosses a thin separating layer of plastic between samples and then goes through the workpiece sample machined with UAT. Each run consisted of three parts: coarsely placed indents (with the distance of 50 μm between two neighbouring indentation points) in the regions distant from the machined surfaces and finely placed indents (with spacing 10 μm) in the direct vicinity of the machined surfaces.

Each indentation point was produced with a constant load of about 10 mN applied to the probe tip. The difference in the residual depth of the indent produced with the probe is automatically recalculated into hardness and the Young's modulus of the corresponding indentation place as it is shown above.

Results of nanoindentation for INCONEL 718 machined with two turning techniques (Figure 13) vividly demonstrate that at distances from the machined surfaces larger than 100 μm the measured levels of nanohardness converge to a relatively low level that should correspond to that for the non-machined (as-delivered) state of the alloy. Sub-surface layers are significantly harder for both techniques, and both the level of hardness and depth of the affected layer are higher for CT.

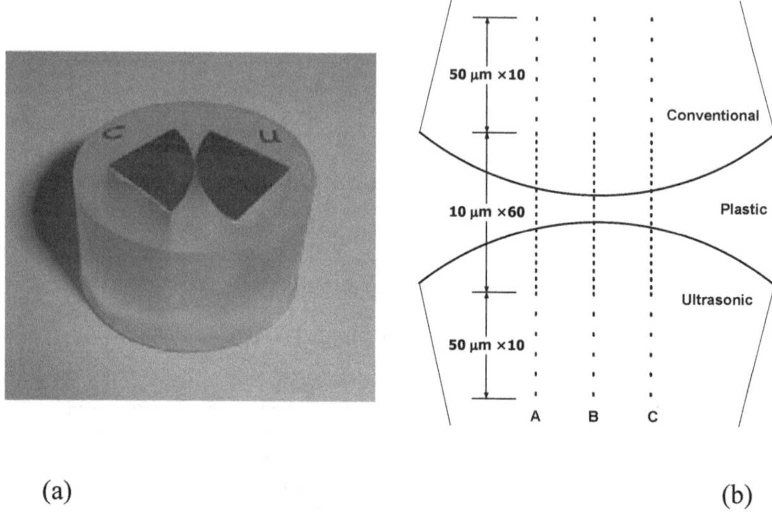

(a) (b)

Figure 12. (a) Tested specimens; (b) schematic of indentation points.

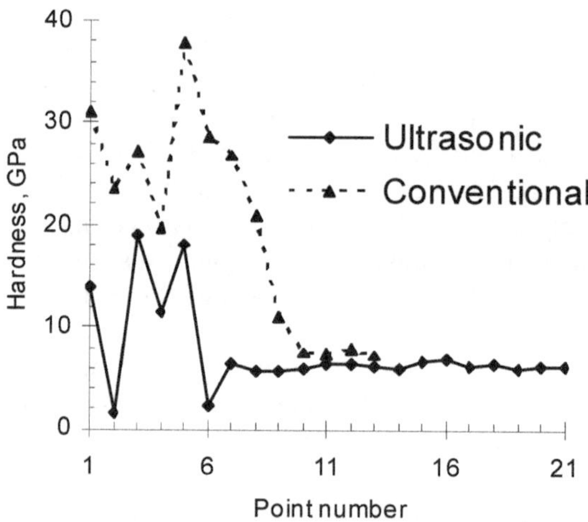

Figure 13. Results of nanoindentation of INCONEL 718 for various machining techniques.

The average width of this layer is 70 % higher for the conventionally machined specimen than for the ultrasonically machined one (85 μm and 50 μm, respectively) while the average hardness of the hardened surface layer for UAT (about 15 GPa) is about 60 % of that for CT (25 GPa). The through-depth scatter in hardness values, i.e. peaks/troughs found in the graph for both CT and UAT machined surfaces (**Error! Reference source not found.**), can be explained by a combination of several factors, most probably, effects of microdefects, second phase/precipitates or carbides/nitrides observed by SEM (Mitrofanov, 2004).

Effect of Porosity in Ceramic Coatings. Ceramic coatings are broadly used as thermal barriers to protect various components from detrimental effects due to high-temperature environment. Usually, such coatings have thickness of several hundreds micrometres and are deposited on the surface of components using various techniques. These deposition techniques result in specific microstructure characterised by oblate grains and a level of porosity, usually in the range from 2% - 8%. The voids have varying dimensions and are randomly distributed in coatings. Such microstructure results in significantly decreased magnitudes of the Young's modulus as compared with those for bulk ceramics. The presence of voids results in local fluctuations in the level of elastic moduli.

A 200 μm-thick plasma-sprayed alumina coating on a metallic substrate was used to study the random character of its mechanical properties. The typical microstructure of this coating that has porosity 1.8% is presented in Figure 14 (Zhao, 2005).

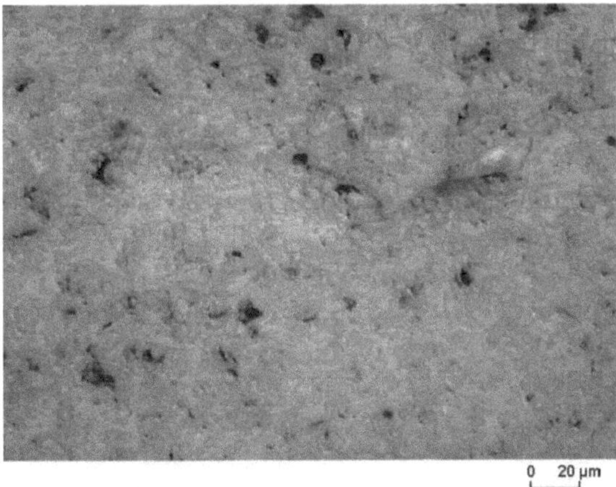

Figure 14. Microstructure of plasma sprayed alumina coating.

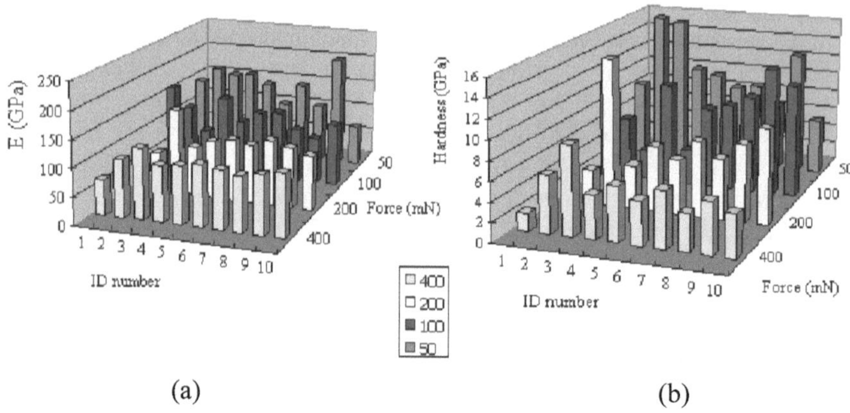

(a) (b)

Figure 15. Young's Modulus (a) and hardness (b) of alumina coating from nanoindentation tests

Nanoindentation tests of this alumina coating were performed using the NanoTest Platform (described above) for various levels of the applied load – from 50 mN to 400 mN. The obtained results for both Young's modulus and hardness (Figure 15) vividly demonstrate a high spatial scatter in these parameters (Zhao, 2005, and Zhao and Silberschmidt, 2006). The analysis shows that they also depend on the level of the applied load: both the average levels of parameters and their scatter decreases with the increase in the indentation load. This can be naturally linked to the effect of the microstructure and the penetration depth.

The lowest force corresponds to a low maximum penetration depth – approx. 550 nm – that is considerably lower than the characteristic size of the microstructural features. Hence, the obtained parameters do not account for the properties reduction due to porosity and rather represent those for the bulk materials. Larger deviations from the average values are linked to indenting various microstructural features, e.g. intra-oblate areas. The increase in the indentation load causes the increase in the maximum penetration depth – more than 2000 nm for 400 mN, and, subsequently, the increased volume of response to indentation. The latter causes, on the one hand, the reduction in the scatter in the measured data due to increased representativeness of the assessed micro-volume. On the other hand, the results for higher loads account for the effect of porosity, which is now a part of the tested volume, on the Young's modulus and hardness (i.e. diminishing their values).

5 Full field measurement techniques

5.1 Introduction

The aim of Section 5 is to introduce a broad picture of some of the major full field techniques currently employed in mechanics of materials and structures, and the use that an engineer can make of information they provide.

Many mechanical and structural engineers are familiar with the ubiquitous strain gauge, which provides an inexpensive and reliable way to measure point-wise surface strains in a wide range of materials. However, when multiple measurements are required, e.g. for mapping a spatial strain distribution on a structural component, full field methods become more suitable. There is currently a plethora of full field measurement techniques that provide displacement or strain distributions for structural parts under load, most of them suitable for static and dynamic studies of mechanical phenomena in objects ranging from nanometres to several meters in size.

Photoelasticity has been developed in the 1930s and is still used to visualize strain/stress fields in transparent materials using the principle of stress-induced birefringence. It mainly applies to flat samples under plane stress, but three-dimensional photoelasticity is also possible either by using frozen stress photoelasticity (achieved by initial stress freezing in the model and then mechanically slicing it) or tomographic photoelasticity (using image processing tomographic reconstruction or light sheet slicing and scattering photoelasticity) (Ramesh, 2000, and Dally and Riley, 1991). The information on the stress field is encoded in a fringe pattern that depends on the phase retardation between two orthogonally polarized light beams propagating through the material. The measured phase retardation is proportional to the difference between the principal stresses on a plane perpendicular to the direction of propagation, and it is also possible to determine the orientation of the principal stresses.

Moiré-based methods have been used for surface strain measurement for more than half a century now and rely on the analysis of fringe patterns that arise when a grating applied or projected onto the object under study is deformed and superimposed (by geometric, digital or interferometric means) to a reference unloaded grating (Bromley, 1956, Brandt, 1967, Durelli and Clark, 1970, Sciammarella and Chawla, 1977, Rastogi et al., 1982, and Post, 1994). The amplitude or phase gratings generally consist of parallel lines or a family of parallel lines orthogonal to each other. Moiré systems are sensitive to displacements perpendicular to the grating lines, with displacement sensitivity being inversely proportional to the grating pitch.

The *Grid method* is based on the replication of a low spatial frequency (circa 10 lines/mm) amplitude grating onto the object under study. Instead of being used to produce a moiré pattern this grating is imaged with a digital camera and lenses that can resolve the lines so that around 5 pixels are used to sample each grating period. Then phase evaluation methods are used to extract the spatial phase before and after deformation of the object, to finally convert the phase changes in displacements.

The advent of the laser in the 1960s brought a formidable development of full field measurement techniques based on *Holographic Interferometry* (Spetzler, 1972, Jones and Wyeks, 1989 and Sirohi, 1993) and then *Speckle Interferometry* (Sirohi, 1993) and *Digital Holography* (Osten, 2007) in the 1970s and 1980s. Two latter techniques took over the former thanks to technological advances in imaging sensors and increased computing power that avoided the wet stage of hologram development and enabled direct numerical analysis. Interferometric techniques measure the optical phase delay between a reference beam and light reflected off the object, with both beams coming from the same coherent source. In the case of optically rough objects, where the roughness is of the order $\lambda/4$, with λ being the wavelength of light, a random interference pattern known as "speckle" is produced, from which it is still possible to extract deterministic phase changes due to object deformation. The optical phase is related to object displacements through the geometry of the setup and the wavelength of the light source used. Due to the short wavelength of visible light (400-750 nm) the displacement sensitivity can be below $\lambda/20 \sim 20$ nm for rough surfaces and around 1 nm or below for smooth specular surfaces.

Speckle Shearing Interferometry belongs to this family of techniques and provides fringe patterns, in which the measured optical phase is proportional to changes in the spatial derivative of the out-of-plane displacements or, in other words, to changes in the slope of the object's surface. In this technique, an image of the object under study is coherently superimposed onto the same image shifted along one specified direction. The sensitivity of the system is proportional to the shift introduced and can be easily adjusted in most setups usually by simply tilting a mirror (Sirohi, 1993). This feature and the fact that it is robust against environmental disturbances such as vibrations and convective currents make it very suitable for damage detection. Composite panels with delaminations, for instance, are usually loaded using vacuum, infrared lamps or even hot air guns.

Another technique, which evolved rapidly aided by advances in imaging technology and computing, was *Digital Image Correlation* (*DIC*), now growing strongly in popularity in the engineering testing community. DIC spun out from the field of photogrammetry, image registration and Particle Image Velocimetry in the 1980s, it is generally (but not necessarily) based on grey scale digital images, triangulation and local correlation, and can provide 3D surface shape and displacement measurements (Popp et al., 1973, Chu et al., 1985, and

Kahnjetter and Chu, 1990). It does not rely on special light sources, as do interferometric techniques, but on the way position information is retrieved from one or multiple cameras viewing the same scene before and after a change in the spatial configuration of the object or flow under study. The sensitivity is usually expressed in fractions of pixels and it can be better than 1/100 of a pixel under certain conditions. The displacement sensitivity therefore depends on the lateral resolution of the imaging system.

Neutron Diffraction and *X-ray Diffraction* can provide strain/stress measurements *within* the volume of materials with a crystalline structure. As a material irradiated with neutrons or x-rays is mechanically loaded, the lattice parameters related to the inter-atomic distances change, which results in shifts of the diffraction peaks. These techniques are usually based on point measurements but full cross sections and volumes can be measured by scanning the measurement gauge volume across the material. Measurements are expensive (£3000/day as at 2005 at the Engin-X facility, Rutherford Appleton Laboratory, UK) as the required high intensity neutrons are generally produced at spallation source facilities or nuclear reactors and x-rays at synchrotron sources, but new technologies such as free electron lasers are leaving the drawing board (Service, 2002) and promise to become the new competitors in the x-ray generation.

The above mentioned techniques constitute just a sample of a much broader collection, enriched by the fact that advances in some of them foster advances in others. For instance, fringe analysis methods, which are mathematical tools to extract phase and displacement information from fringe patterns wherever the fringes come from (photoelasticity, moiré, interferometry, etc.) are common to different techniques and have become an area of research on its own (Osten, 2005). Another important refinement of phase measurement methods is their ability to produce enough data, by using phase shifting methods (Huntley, 2001), to extract both the absolute value and the sign of the phase and therefore of the associated displacement field.

5.2 Experimental validation of numerical predictions

There are a number of ways in which full field measurements assist and advance mechanical engineering research. The straightforward use of 2D distributions of displacements and strains is to identify weak points and regions of strain/stress concentration in structural components, as well as regions of low strains/stresses that could be removed to reduce the material's use. This way of feeding back information to design engineers has its drawbacks, generally linked to time, costs and the need for experimental facilities. For these reasons, the role of numerical modelling has become more and more important and predictions obtained using the Finite Element Method (FEM) can now drive the design stage. However, unless engineers fully validate their FEM models against experimental measurements (to check their assumptions and geometrical and material properties of the elements used), they will not be in a position to trust their predictions. This is another role of full-field techniques.

In processed metals, residual stresses are frequently produced by localised plastic deformation and in polymers a similar effect may occur by localised visco-elastic or visco-plastic deformation. Differential expansion or contraction in a bonded structure is also a major source of residual stress. In an adhesively bonded joint differential expansion can be caused by changes due to chemical reactions, temperature and moisture (or other absorbed solvent) content, leading to curing, thermal or hygroscopic stresses, respectively. In most occasions it is not possible to distinguish between each other, but their consequence is that there is some residual stress in the joint components before any mechanical load is applied.

The following example aims to illustrate how full-field measurement methods can be used to validate FEM models by straightforward comparison of measured and predicted results and without using any adjusting parameters (Jumbo et al., 2004). On the one hand, residual strains and total internal strains were measured in the aluminium (Al) adherend of an aluminium/carbon fibre reinforced polymer (Al/CFRP) double lap shear joints using neutron diffraction.

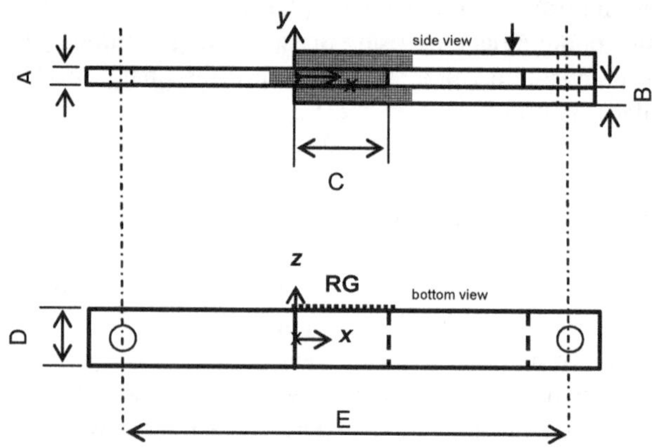

Figure 16. Schematic of a double lab joint. RG represents the reflection grating replicated on it for moiré interferometry measurements.

On the other hand, full field surface strain owing to the mechanical stresses alone was measured using *Moiré Interferometry*. Experimental results were compared against FEM predictions obtained by using the known geometry of the joints, independently measured mechanical properties and the applied loads/thermal gradients during the curing process.

Validation of residual stress predictions with neutron diffraction. Figure 16 shows a double-lap adhesively bonded joint. The outer adherends material used was unclad 7075-T6 aluminium alloy and the middle adherend was uni-axial IM7/8552 CFRP. The adhesive used in the experiments was FM73 Cytec, 1998). The dimensions of the sample were as follows: A = 3 mm, B = 3 mm, C = 25 mm, D = 25 mm and E = 250 mm.

The Al/CFRP double-lap (DLS) joint full geometry was modelled with 44840 eight-node, full integration 3-D brick elements. MSC.Marc element 7 with assumed strain formulation was used for the adhesive and aluminium substrate while element 149, which is a 3-D, eight-node composite brick element, was used for the CFRP. In numerical simulations, the load was applied to the joint via a pin, as in the experiments, using a rigid-to-deformable body-contact analysis, with the pin as the rigid body and the joint as the deformable body. For analysis of the thermal residual stresses, an initial condition of 120°C was applied to the models and a uniform cooling rate was applied for a change in temperature of 100°C. A quasi-static load of 5 kN was applied along the axis of the specimen for measuring surface strains on the grating region.

Figure 17 shows the FEM predictions and Neutron Diffraction measurements of the longitudinal (ε_{xx}) and peel (ε_{yy}) thermal residual strains in the Al/CFRP joint for different positions along the middle line of the outer aluminium adherends. Within the accuracy limits, good agreement is observed between the measured and predicted strains in four out of six points measured in Figure 17 (a) and in four out of five points in Figure 17b. The maximum predicted longitudinal residual strain of 4.5×10^{-3} agrees well with the measured values at the middle of the joint length (x = 6 mm). These are well in excess of the maximum values of circa 10^{-5} strain predicted and 2×10^{-4} measured for the aluminium double-lap joint, and confirms that hybrid joints with a high temperature curing adhesive will have considerable residual stresses before the application of any load. These stresses should therefore be taken into consideration before any realistic modelling of applied load can take place.

Validation of surface strain predictions with moiré interferometry. In Moiré Interferometry, symmetric coherent illumination beams intersect at an angle onto the sample in order to generate a virtual grating with twice the spatial frequency of a diffraction grating replicated onto the sample. A schematic view of the moiré interferometer with horizontal (x-axis) sensitivity is shown in Figure 18.

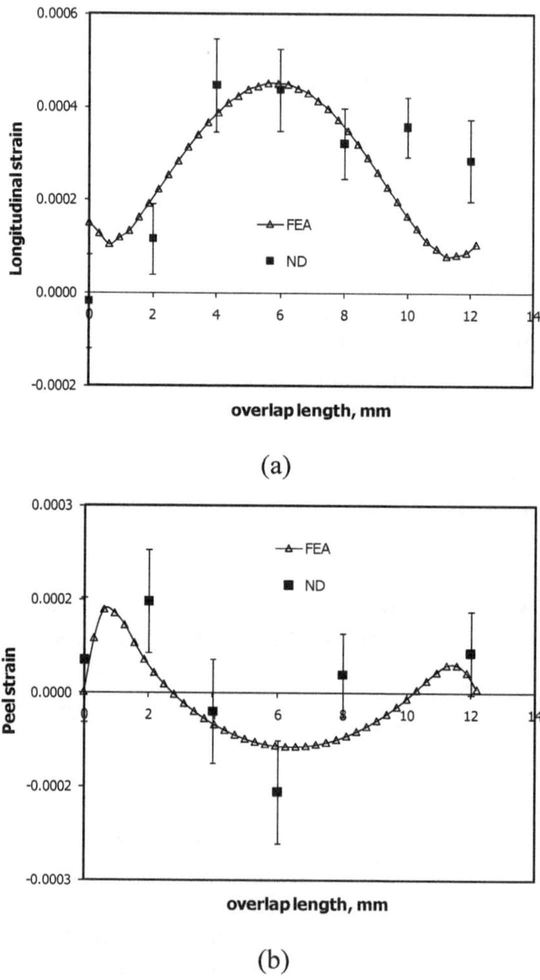

Figure 17. Neutron diffraction results and FEA predictions of longitudinal and peel strain for an Al/CFRP joint.

Illumination angles of 18.6° were used for sample gratings of 600 lines/mm

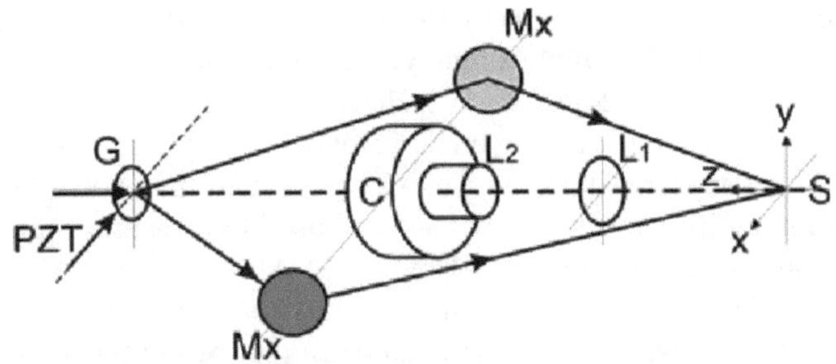

Figure 18. Moiré interferometer showing: grating beam splitter G; piezoelectric transducer PZT for phase shifting; steering mirrors M_x; sample with replicated grating S; Field lens L_1; Imaging lens L_2 and camera C.

and an illumination wavelength of $\lambda = 532$ nm. Quantitative displacement fields were obtained across the replicated grating by evaluating the optical phase change between the reference and loaded states of the sample. Phase changes of 2π are produced for displacements of half the pitch of the sample grating, and are therefore equivalent to an in-plane displacement of 0.83 μm along the x-axis (see Figures 16 and 18). While displacement fields are experimentally measured, strain is evaluated by numerical differentiation of the displacement fields.

Phase shifting was implemented by moving the grating beam splitter along the x-axis with a piezoelectric transducer. This provided the magnitude and sign of the displacement field $u(x, y)$ (u being the sample's displacements along the x-axis), which was numerically differentiated to obtain the distribution of strain $\varepsilon_{xx}(x, y)$ shown in Figure 19a. Figure 19b shows the FEM predicted value of $\varepsilon_{xx}(x, y)$.

Figure 20 shows cross sections $\varepsilon_{xx}(x,0)$, $\varepsilon_{xx}(x,-3.4)$ and $\varepsilon_{xx}(x,3.4)$ of the longitudinal strain distribution $\varepsilon_{xx}(x, y)$ as measured with moiré interferometry and predicted with an FEM model assuming uniform loading. This study has shown that in a joint with dissimilar adherends such as the Al/CFRP joint, the predicted and measured strains are in close agreement in their spatial distribution and magnitude. Within the accuracy limits of measurements using neutron diffraction, the experimental work has confirmed the FEM predictions and shown that significant strains exist in an Al/CFRP DLJ before the application of any load. It is therefore important that these strains are considered in any prediction of in-service joint behaviour.

Validation of surface strain predictions within the adhesive layer using high magnification moiré interferometry. Reliable predictions of strain/stress distributions in the adherends of adhesive joints can be obtained with appropriately validated FEM models. Bonded joints, however, fail due to fracture of the adhesive line, which is the weakest point in the joint. High-magnification moiré interferometry has been used to measure strain distributions within the adhesive line in the fillet region of Al/Al and Al/CFRP joints under uniaxial load along the *x*-axis. A dual sensitivity interferometer similar to the one shown in Figure 18 was used but with an extra pair of illumination beams in the *yz* plane to provide sensitivity along the *y*-axis. This requires crossed lines gratings replicated onto the region of interest, which in this case was of about 1.7 mm × 0.8 mm. Fields of displacements $u(x, y)$ and $v(x, y)$ (along the *x*- and *y*-axis, respectively) were measured, from which all the components ε_{xx}, ε_{yy} and γ_{xy} of the 2D engineering strain tensor were obtained through numerical differentiation. The illumination angles, grating frequency and illumination wavelength were the same as those used in the system described in the previous section, leading to the same displacement sensitivity of 0.83 μm for every 2π rad phase change.

The FEM package MSC Marc was used to model the bonded joint in the small region of interest of the adhesive line in the fillet area. In order to reduce the complexity of the model and computational demands, the model was simplified as much as possible without compromising the results, as discussed in (Jumbo et al., 2004). The measured adhesive fillets were included and the full

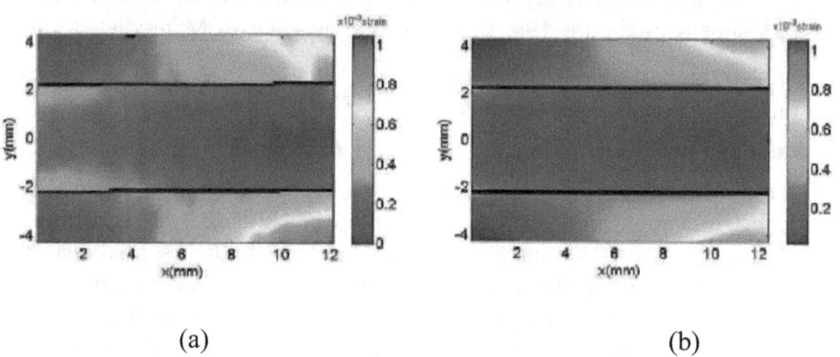

Figure 19. Comparison between moiré interferometry results (a) and FEM predictions (b) of longitudinal strain ε_{xx} obtained at the surface of the CFRP/Al joint under a tensile load of 5 kN.

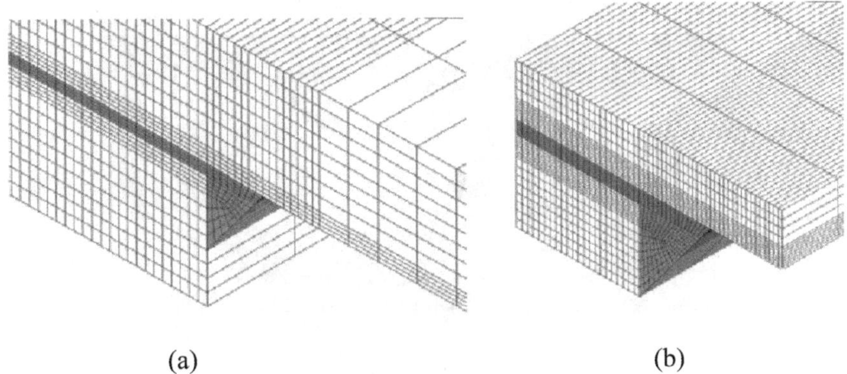

Figure 20. FEM mesh for Al/Al double lap joint: (a) global mesh; (b) local refinement

joint geometry was modelled for each joint to account for manufacturing misalignment in individual samples.

The joints were modelled using 8 node full brick 3D elements with assumed

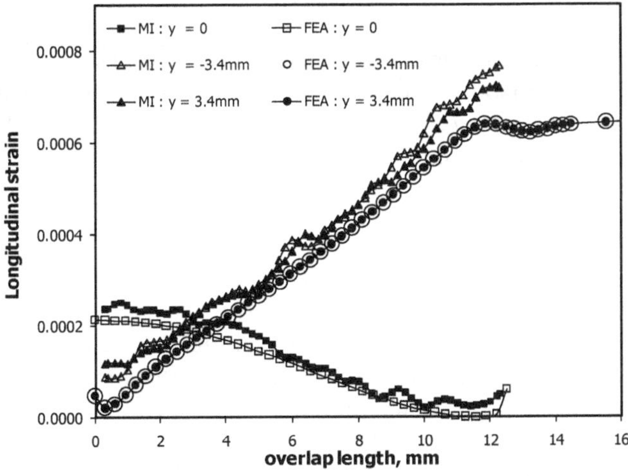

Figure 21. Comparison of moiré interferometry and FEM longitudinal strain distributions across the overlap length of the DLJ.

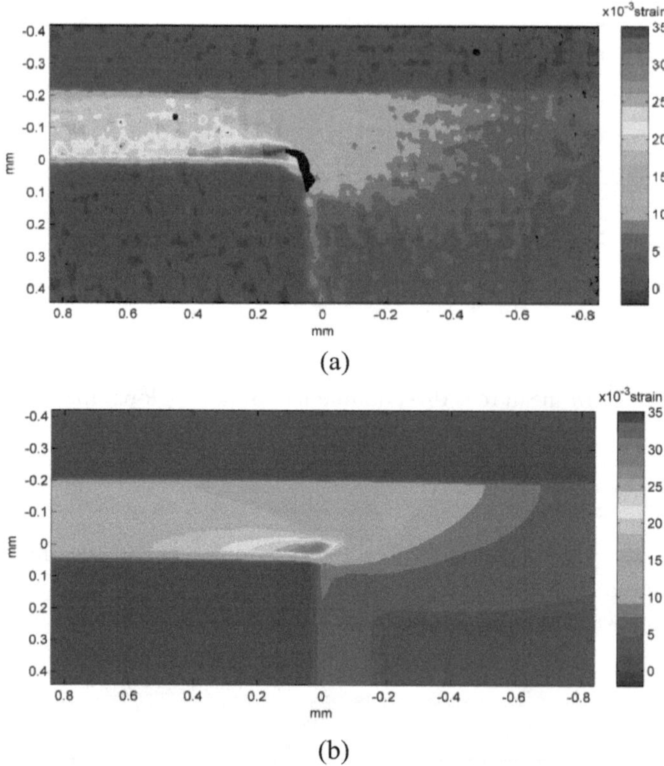

Figure 22. Engineering shear strain distribution $\gamma_{xy}(x, y)$ in adhesive layer of Al/Al double lap joint under 7 kN tensile load: (a) measured with high-magnification moiré interferometry; (b) predicted with FEM model, which includes local refinement and plastic deformations.

strain formulation for the adhesive and aluminium adherends and 8 node 3D composite brick elements for CFRP adherends.

Owing to the high resolution of strains required over a small area, a sub-modelling approach, termed structural zooming in MSC Marc (MSC, 2005), was used to improve the strain results in an efficient way. The structural zooming analysis involves a global model to derive the boundary conditions for a highly refined local model to obtain refined results in the region of interest. This procedure can be repeated as many times as desired and any local analysis can be the global analysis for the next level of refinement. Figures 21a and b show respectively the global mesh and the local mesh refinement used in the fillet area for a local model. For each joint, a 7 kN load was applied along the x-axis and measured mechanical properties were used for the adherends and the adhesive (Jumbo et al., 2004).

The adhesive line thickness was around 200 μm. Figure 22a shows the engineering shear strain γ_{xy} measured in one of the adhesive lines in the fillet area of an Al/Al double lap joint under a 7 kN uniaxial load. Figure 22b shows the FEM prediction of the same strain field under the same loading conditions, using local refinement and considering plastic deformation.

There is a great similarity between the measurements and the FEM predictions, both of them showing a region of shear strain concentration within the adhesive layer and next to the outer adherend. An optical micrograph in Figure 23 shows that this is indeed the region at which the crack initiates.

5.3. Damage detection and characterization

Damage detection is another common application of full field techniques and usually requires the object to be loaded by mechanical, thermal, or other means that will excite a mechanical response of the damage, which is different to the response of the undamaged material. Photoelasticity using coating methods, moiré-, speckle- and shearing- interferometry can effectively provide the necessary mapping of strain/stress fields in a broad range of materials such as glass, metals, ceramics, composites or polymers. Figure 24 illustrates the principle in the particular case of a carbon fibre reinforced polymer (CFRP)

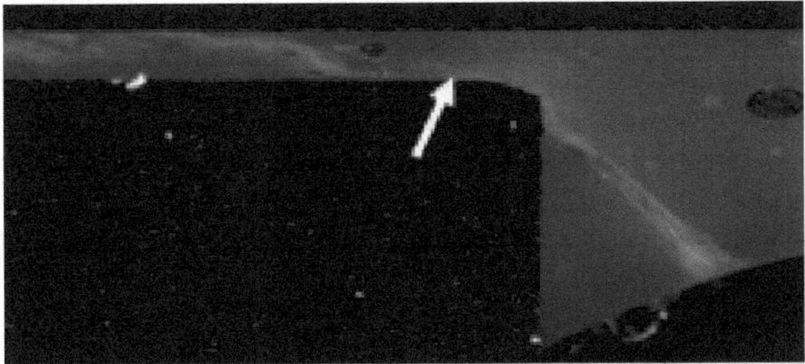

Figure 23. Crack through the fillet and adhesive line, initiated at the region of stress concentration predicted by the models and measured with moiré interferometry (arrow).

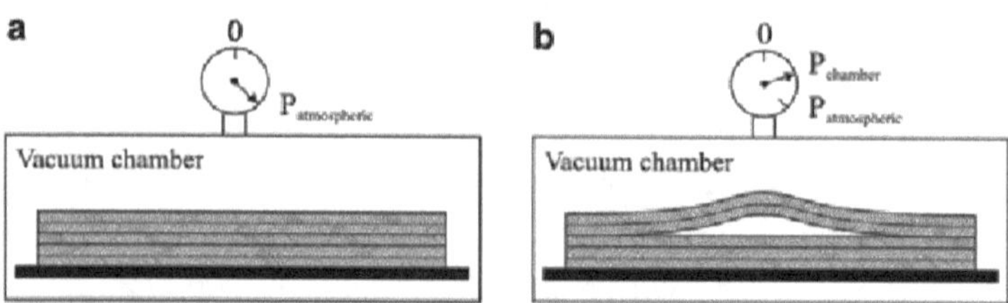

Figure 24. Laminate with internal delaminations in a vacuum chamber before (a) and after (b) reducing the pressure of the vacuum chamber that houses the sample.

panel with an artificial delamination made by embedding a thin Teflon film under the first ply. The panel was subjected to a change in environmental pressure, and temporal phase shifting speckle interferometry was used to measure the time-resolved evolution of the deformation as the pressure changed in a chamber (Maranon et al., 2007, and Davila et al., 2003). Figure 25 shows a schematic of the optical setup used. The displacement field due to surface out-of-plane deformation was measured and it is shown in Figure 26. In this case, with the panel containing a circular artificial delamination, the same experimental setup was used to detect real delaminations caused by impact and to study various improvements of the data processing algorithms.

Low-noise, high-sensitivity full field measurements like these can be used to *characterize* the damage. The *inverse* problem of characterization can be formulated as an optimization problem, in which the shape, lateral position (along the x- and y-axes) and depth of the delamination are estimated by using measured displacements produced by a known load. The more familiar *direct* problem consists of calculating the displacement field of the panel surface when the mechanical properties of the panel, the depth, position and size of the delaminations and the applied loads are all known. The procedure, illustrated in Figure 27, is as follows:

0- A panel with an unknown delamination is vacuum-loaded and the displacement field is measured.
1- An edge detector operator is used on the measured displacement distribution to find the edges of the delamination.
2- An ellipse is fitted to the detected edge in order to reduce the parameters that describe the shape (data reduction).
3- Central geometric moments are evaluated from the displacement measurements. This is another data reduction stage in which a small

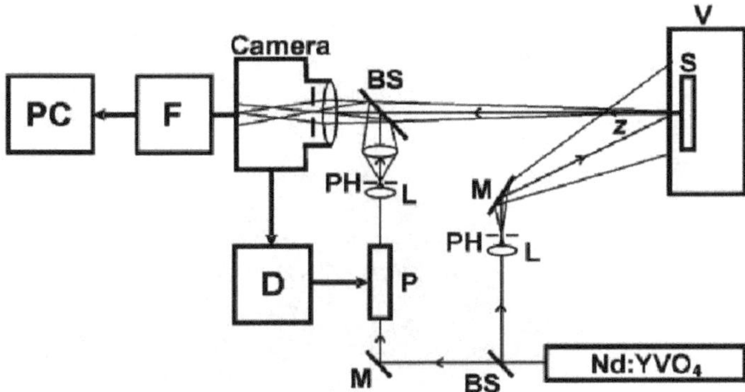

Figure 25. Temporal phase shifting optical setup for time-resolved speckle interferometry showing Nd:YVO_4 laser, beam splitter BS, mirrors M, phase modulator P for phase shifting, phase modulator driver D, lenses L, pin hole spatial filters PH, frame grabber F, personal computer PC, sample S and vacuum chamber V.

number of parameters are found to describe the deformation. Even though this stage is not strictly necessary, it speeds up the solution of the problem without compromising the physical significance of the data.

4- A finite-element model is constructed based on known material properties, geometry of the panel, load applied and the parameters identified in step 2 and an initial guess of the delamination depth and internal pressure.

5- Central geometric moments are evaluated for the predicted FEM displacements.

6- Central geometric moments evaluated from measurements and from FEM predictions are compared and a cost function is evaluated. The FEM model is updated with new parameters generated by using an adaptive Genetic Algorithm (GA). This process is repeated until the cost function is minimized, at which point the delamination parameters are obtained.

By following this methodology, Maranon et al. (2007) found that the planar location of delaminations was predicted with accuracy greater than 95% in nine out of ten samples studied. Even though the depth of the delaminations was predicted correctly only 50% of the times, in those cases where the delamination

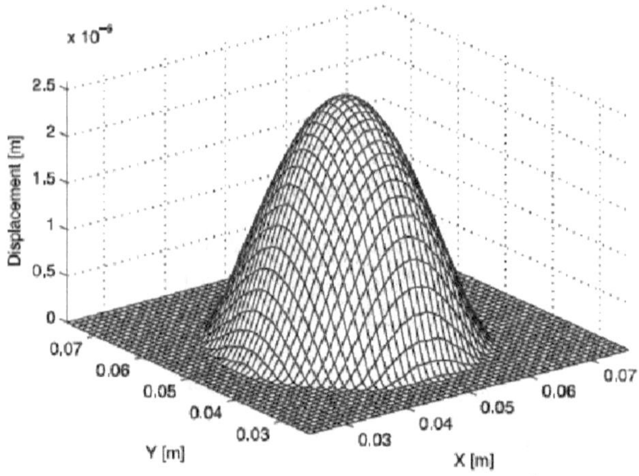

Figure 26. Displacement field due to embedded artificial delamination in CFRP measured with temporal phase shifting speckle interferometry.

was not predicted accurately, the identification methodology reported depths that were close to the real value by just one interlaminar layer above or below of the correct location. Results for the pressure of the air trapped inside of the delamination are inconclusive, as there are no experimental measurements or analytical calculations to verify their correctness.

One interesting conclusion from this work was that the problem of delamination depth characterization is not well posed and multiple solutions may exist that result in the same values of the cost function. In other words, different delaminations can result in the same response: a big and deep delamination can result in the same (in the sense that surface measurements alone cannot tell the difference) displacement distribution than a smaller and shallower delamination.

In Section 5.5 we will discuss one way to overcome this problem, which consists in measuring the displacements in the whole volume of the material.

5.4. Identification of material constitutive parameters

Once the displacement fields of an object under known loads have been measured, they can also be used to find the constitutive parameters of the material. In the general case where the stress/strain fields are heterogeneous due

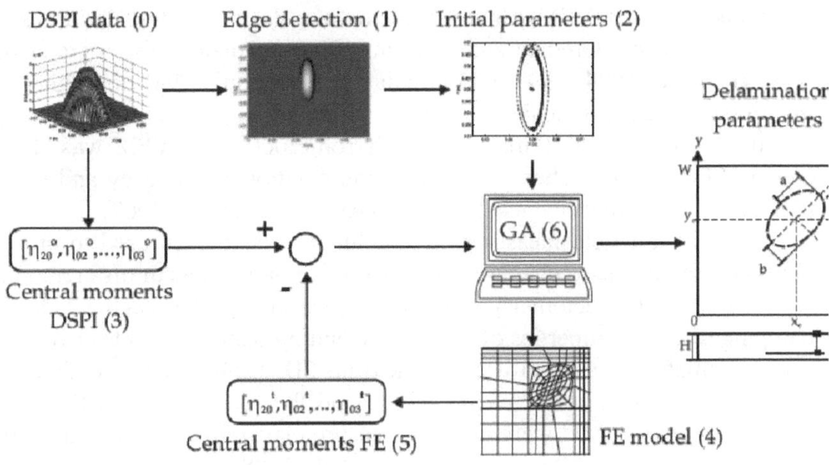

Figure 27. Methodology for characterization of delaminations in composite panels using Finite Element Updating (FEMU) and Genetic Algorithms.

to boundary conditions, specimen's geometry, localized plasticity or damage, the inverse problem can only be solved if full-field data is available. Finite Element Method Updating (FEMU) (Pedersen and Frederiksen, 1992) and the Virtual Fields Method (VFM) are two approaches currently used to obtain the constitutive parameters from 2D full-field strain measurements. In the former, the direct problem is solved by means of the Finite Element Method as illustrated in Section 5.3 by using initial guesses of the unknown parameters. Simulations are performed iteratively until the displacements computed at various nodes of the mesh match their experimental counterparts. In the VFM, for measured strain fields, the stress fields are expressed as a parameterised function of the unknown constitutive parameters. The principle of virtual work is then applied so that the stress fields verify the global equilibrium of the structure. Finally, the use of several virtual fields yields a system of equations that involve the unknown parameters leading to the solution of the problem (Grediac, 1989, Pierron, 2000, and Grediac et al., 2006).

The VFM has matured in the last two decades to the point, in which a more robust and efficient characterization is performed as compared to FEMU. By using the simulated 2D surface displacement/strain data, Grediac and co-workers illustrated how the VFM can be used to find elastic bending rigidities of thin anisotropic plates (Grediac and Vautrin, 1990), in-plane anisotropic stiffnesses (Grediac et al., 1994, 1999, and Grediac and Pierron, 1998), through-thickness moduli of thick composite panels (Grediac et al., 2001) and elasto-plastic constitutive parameters (Grediac and Pierron, 2006). The VFM was also tested against FEMU approaches in terms of computational efficiency and robustness to noise in the displacements/strain data (Avril and Pierron, 2007).

Different full field measurements techniques have been used to provide the necessary input data to characterize constitutive parameters of different materials with the VFM. Deflectometry was used to measure 2D displacements to find the damping material properties of isotropic vibrating plates (Giraudeau et al., 2006). The grid method was also used to measure 2D displacements to find stiffness parameters of glass epoxy beams (Avril and Pierron, 2007) and Digital Image Correlation and FEMU were used to measure elasto-visco-plastic parameters on steel specimens at high strain rates (Avril et al., 2008).

Even though 2D surface measurements can be sufficient to characterize material's constitutive parameters such as the ones mentioned above, in order to map the spatial distributions of those parameters full field 3D measurements are required. This is relevant when heterogeneous materials are studied at a scale, in which different constituents are resolved, e.g. fibres and the matrix in a composite material. In the next section we discuss some exciting new developments in experimental techniques that, together with VFM or FEMU approaches, could enable spatially resolved material characterization in partially scattering materials.

5.5. Towards the development of suitable full-field measurement techniques for spatially resolved identification of constitutive parameters

Semitransparent scattering materials are widely encountered in industrial applications (adhesives, polymers, composite materials) and biological systems (tissues such as skin, cartilages, cornea) and their function is often critical in the performance and structural integrity of the system they belong to. There has been a long history of research into the mechanical and functional behaviour of these materials, but in many cases they are not fully understood due to

(i) the inherent complexity of the materials that cannot always be modelled with continuum mechanics assumptions, as in the case of many composite materials and biological tissues,

(ii) the limitations of current experimental techniques to measure the distribution of the constitutive parameters of the materials *within* their volume with enough sensitivity and spatial resolution, in order to relax the need for more simplistic bulk properties assumptions.

As discussed in previous Section, displacement fields can be used to evaluate the constitutive parameters of an anisotropic material by solving an inverse problem. In general it requires the measurement of three-component displacement fields (to over-determine the problem) under a set of known boundary conditions, and the geometry of the sample (Avril and Pierron, 2007). The inverse problem is usually posed as identification of the most representative components of the material's constitutive matrix. Generally, this is due to insufficient spatial resolution and/or incomplete multi-component displacement mapping, as is the case for experimental methods that only provide surface displacement fields. Therefore, techniques that allow us to measure multi-component displacement fields inside the volume of the material will enable us to determine non-uniform distributions of constitutive parameters, which is essential to accurately predict the mechanical behavior.

A broad range of methods to measure *internal* structure and displacement fields have been developed in the last few decades such as neutron and X-ray diffraction, Photoelastic Tomography (PT) (Abe et al., 1986, and Abe et al., 2004), Phase Contrast Magnetic Resonance Imaging (PCMRI) (Steele et al., 2000, and Draney et al., 2002), and 3-D Digital Image Correlation using data acquired with X-ray Computed Tomography (XCT) (Bay, 1999) and Optical Coherence Tomography (OCT) (Schmitt, 1998, and Fercher et al., 2003). Each technique has a restricted range of materials to which it can be applied: PT, for example, is suitable only for materials that exhibit photo-elasticity; PCMRI requires significant water or fat content in the sample, and neutron diffraction relies on the crystalline structure of the material. For many technologically- and medically-important materials the existing techniques are often either non-applicable, have insufficient spatial resolution or are too insensitive.

OCT is an exciting technique that provides depth-resolved microstructure images primarily for medical applications. It is based on a Michelson interferometer and a low temporal coherence broadband source and is usually implemented in the time domain, in which case the reference mirror is scanned to provide cross-sections of the sample. It can also be realised in the spectral domain (SOCT) where all the information for a 'slice' inside the material is registered simultaneously by using a spectrometer, an area photodetector array and no scanning devices. A 2D interferogram is recorded with depth encoded as spatial frequency along the wavelength axis of the spectrometer, rather than as a function of time (Dressel et al., 1992). The microstructure in the 10^{-3}-10^{-6} m range is then extracted from the spectral magnitude of the Fourier transform along the wave-number axis.

Although digital correlation methods can be used to quantify displacement fields from OCT microstructure images obtained before and after deformation, the displacement sensitivity would be limited by the depth resolution of typically 10 μm or more, with 1 μm only achievable with ultra-high resolution systems. This constrains the ability to detect and quantify deformation fields due to small loads (mechanical, thermal or chemical) even for compliant materials. It was recently demonstrated that optical phase information can be extracted from OCT data to measure displacements with a sensitivity of order 10 nm, i.e. some 2-3 orders of magnitude better than the intrinsic depth resolution. Three main different methods that make use of phase information to measure depth-resolved displacements have been used and these will be referred here as Phase-Contrast (PC) methods (Vakoc et al., 2005). These are:

1) Tilt Scanning Interferometry (TSI);
2) Wavelength Scanning Interferometry (WSI);
3) Phase Contrast Spectral Optical Coherence Tomography (PC SOCT)

and will be briefly described below. Doppler OCT, proposed in 1997, also makes use of the optical phase information for velocity mapping of e.g. retinal blood flow (Chen et al., 1997).

Measurement of depth-resolved displacement fields in semitransparent materials using optical phase information. TSI is realised by tilting the illumination wavefront from a monochromatic source whilst a sequence of interferograms records depth-encoded temporal carrier signals (Ruiz et al., 2006). TSI (Figure 28b) can be adapted to any laser source and controlled tilting of the wavefront is technically much easier than tuning the wavelength of a laser as in Wavelength Scanning Interferometry (WSI) (Figure 28a), a sequential version of spectral OCT that can also deliver depth-resolved displacement fields (Ruiz et al., 2004, 2005). For example, closed loop piezo-electric tilt stages are now available with sub-μrad resolution that can scan a beam through 100 mrad in a fraction of a second. At an incidence angle of 45°, and a wavelength of 532 nm the effective depth resolution in TSI is ~30 μm. To achieve the same resolution, a WSI system would need a tuning range of ~30 nm, requiring mode-hop-free dye or Ti:sapphire lasers, which operate on multiple longitudinal modes with the potential for major speckle decorrelation.

TSI is appropriate to study static problems and it can provide *3-D distributions of all displacement components* within a scattering material, with high sensitivity and high spatial resolution. As an example of this approach, depth-resolved displacement measurements with sensitivity to out-of-plane and one in-plane displacement component have been made within a point-loaded polyester resin beam (Ruiz et al., 2006). The top row of Figure 29 shows the depth-resolved in-plane wrapped (2π modulus) phase change (proportional to in-

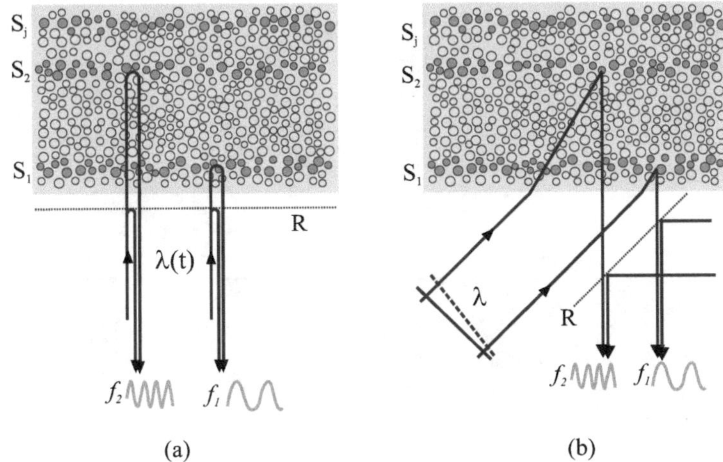

Figure 28. Generation of depth-encoding frequency shifts by: (a) tuning the wavelength of the light source through a narrow range (Wavelength Scanning Interferometry); (b) tilting a monochromatic wavefront (Tilt Scanning Interferometry).

plane displacements) starting at the object surface $z = 0$ mm (left) in steps of 1.74 mm down to $z = 5.22$ mm (right). The wrapped out-of-plane phase change distribution for the same depth slices are shown in the bottom row of Figure 29.

PC SOCT is fundamentally analogous to WSI, the main difference being that in PC SOCT a low cost broadband source is used and the interference signal is present along the wavelength axis of a spectrometer detector, rather than along the time axis of a sequence of frames, as in WSI. The key feature of PC SOCT is that it is based on "single shot", which enables fast acquisition of instantaneous deformation states to study dynamic events and follow the temporal evolution of displacement and strain fields with nanometre sensitivity within a slice in the sample. It provides 2D cross sections of the material, usually perpendicular to , the surface of the sample. Out of plane sensitivity is straightforward to measure and in-plane sensitivity has been recently achieved. Depth-resolved displacements were measured by means of PC SOCT achieving a depth resolution of circa 30 microns and a displacement sensitivity of circa 30 nm, limited by the phase noise (De la Torre-Ibarra et al., 2006). PC SOCT has exciting practical advantages over WSI and the potential to expand the capabilities of SOCT to map displacement and strain fields with nanometre

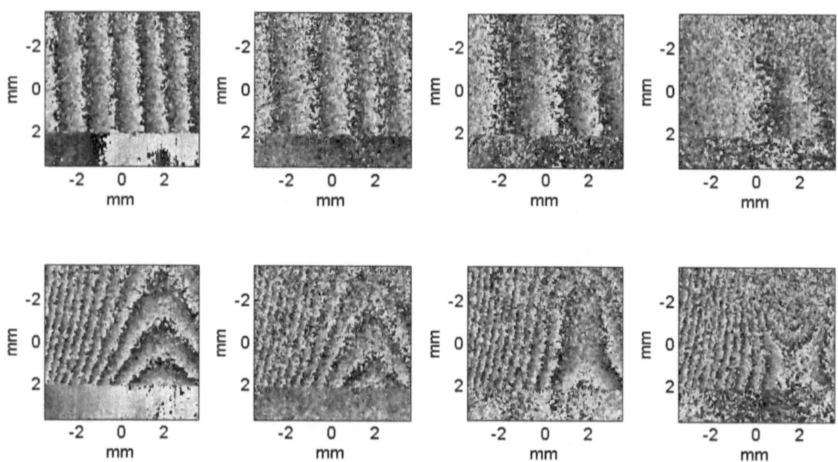

Figure 29. In-plane (top row) and out-of-plane (bottom row) wrapped phase-change distribution for different slices within a simply supported polyester resin beam under pin loading (coordinate of pin is approximately (2, -2). Black represents -π and white +π.

sensitivity within the microstructure of *in-vivo* samples.

Polarization Sensitive Optical Coherence Tomography (PSOCT) is another powerful new technique which is finding interesting applications in mechanics of materials (de Boer et al., 1998, de Boer and Milner, 2002, Yasuno et al., 2003, Strąkowski et al., 2008, Stifter et al., 2003, 2007, 2008, Wiesauer et al., 2005, 2006, 2007, and Stifter, 2007). It combines depth-resolved imaging capabilities of OCT with photoelasticity principles in such a way that the phase retardation of orthogonal polarization components is measured at each point in a slice through the material. As this retardation is proportional to the difference of principal stresses in a plane perpendicular to the observation direction, it is possible to map depth-resolved stresses. It would not be surprising to see a merging of polarization-sensitive with phase contrast OCT methods, to combine their working principles and different sensitivities in a single technique.

Bibliography

Abaqus Version 6.4 Users Manual. Abaqus, Inc., Rhode Island; 2003.

T. Abe, Y. Mitsunaga, and H. Koga. Photoelastic computer-tomography - a novel measurement method for axial residual-stress profile in optical fibers, *Journal of the Optical Society of America A*, 3:133-138, 1986.

H. K. Aben, A. Errapart, L. Ainola, and J. Anton. Photoelastic tomography for residual stress measurement in glass. In *Optical Metrology in Production Engineering*, pages 1-11, SPIE-INT, 2004.

N. Ahmed, A. V. Mitrofanov, V. I. Babitsky, and V.V. Silberschmidt. Analysis of material response to ultrasonic vibration loading in turning Inconel 718. *Materials Science and Engineering A*, 424:318–325, 2006.

K. Alam. Private communication, 2008.

S. Avril, F. Pierron, M. A. Sutton, and J. Yan. Identification of elasto-visco-plastic parameters and characterization of Lüders behavior using digital image correlation and the virtual fields method. *Mechanics of Materials*, 40:729-742, 2008.

V. I. Babitsky, A. V. Mitrofanov, and V. V. Silberschmidt. Ultrasonically assisted turning of aviation materials: simulations and experimental study. *Ultrasonics*, 42:81-86, 2004.

B. K. Bay, T. S. Smith, D. P. Fyhrie, and M. Saad. Digital volume correlation: three-dimensional strain mapping using X-ray tomography. *Experimental Mechanics*, 39:217–226, 1999.

M. Bocciarelli, G. Bolzon, and G. Maier. Parameter identification in anisotropic elastoplasticity by indentation and imprint mapping. *Mechanics of Materials*, 37:855-868, 2005.

J. F. de Boer, S. M. Srinivas, A. Malekafzali, Z. Chen ,and J. S. Nelson. Imaging thermally damaged tissue by polarization sensitive optical coherence tomography. *Optics Express*, 3:212-218, 1998.

J. F. de Boer and T. E. Milner, Review of polarization sensitive optical coherence tomography and Stokes vector determination. *Journal of Biomedical Optics*, 7:359-371, 2002.

F. M. Borodich and L. M. Keer. Contact problems and depth-sensing nanoindentation for frictionless and frictional boundary conditions. *International Journal of Solids and Structures*, 41:2479-2499, 2004.

G. B. Brandt. Hologram-Moiré Interferometry for transparent objects. Applied Optics, 6:1535-1540, 1967.

R. H. Bromley. 2-Dimensional strain measurement by moiré. *Proceedings of the Physical Society of London Section B*, 69:373-381, 1956.

S .I. Bulychev, V. P. Alekhin, M. Kh. Shorshorov, A. P. Ternovskii, and G. D. Shnyrev. Determination of Young's modulus according to indentation diagram. *Ind. Lab.*, 41:1409–1412, 1975.

S. I. Bulychev, V. P. Alekhin, M. Kh. Shorshorov and A. P. Ternovskii. Mechanical properties of materials studied from kinetic diagrams of load versus depth of impression during microimpression. *Strength Mater.*,

8:1084–1089, 1976.

Z. Chen, T. E. Milner, D. Dave, and J. S. Nelson. Optical Doppler tomographic imaging of fluid flow velocity in highly scattering media. *Optics Letters*, 22:64-66, 1997.

R. M. Christensen. *Theory of Viscoelasticity: An Introduction*. Academic Press, 1971.

T. C. Chu, W. F. Ranson, M. A. Sutton, and W. H. Peters. Applications of digital-image-correlation techniques to experimental mechanics. *Experimental Mechanics*, 25:232-244, 1985.

Cytec. FM73 Toughened Epoxy Film. [Electronic Manual], 1998. Available from: http://www.cytec.com.

J. W. Dally and W. F. Riley. *Experimental Stress Analysis*. McGraw-Hill, 1991.

A. Davila, P. D. Ruiz, G. H. Kaufmann, and J.M. Huntley. Measurement of sub-surface delaminations in carbon fibre composites using high-speed phase-shifted speckle interferometry and temporal phase unwrapping. *Optics and Lasers in Engineering*, 40:447-458, 2003.

M. H. De la Torre-Ibarra, P. D. Ruiz, and J. M. Huntley. Double-shot depth-resolved displacement field measurement using phase-contrast spectral optical coherence tomography. *Optics Express*, 14:9643-9656, 2006.

M. T. Draney, R. J. Herfkens, T. J. R. Hughes, N. J. Pelc, K. L. Wedding, C. K. Zarins, and C. A. Taylor. Quantification of vessel wall cyclic strain using cine phase contrast magnetic resonance imaging. *Annals of Biomedical Engineering*, 30:1033–1045, 2002.

T. Dresel, G. Hausler, and H. Venzke. 3-dimensional sensing of rough surfaces by coherence radar. *Applied Optics*, 31:919-925, 1992.

A. J. Durelli and J. A. Clark. Moiré Interferometry with embedded grids - effect of optical refraction – Discussion. *Experimental Mechanics*, 10:175, 1970.

A. F. Fercher, W. Drexler, C. K. Hitzenberger, and T. Lasser. Optical coherence tomography - principles and applications. *Reports on Progress in Physics*, 66:239-303, 2003.

J. S. Field and M. V. Swain. Determining the mechanical properties of small volumes of material from submicrometer spherical indentations. *Journal of Materials Research*, 10:101-112, 1995.

A. C. Fischer-Cripps. *Nanoindentation*. Springer-Verlag, 2nd ed., 2004.

A. Giraudeau, B. Guo, and F. Pierron. Stiffness and damping identification from full field measurements on vibrating plates. *Experimental Mechanics*, 46:777-787, 2006.

J. Gong. *Microstructural Features and Mechanical Behaviour of Lead Free Solders for Microelectronic Packaging*. PhD Thesis, Loughborough University, 2007.

J. Gong, C. Liu, P. P. Conway, and V. V. Silberschmidt. Grain features of

SnAgCu solder and their effect on mechanical behaviour of microjoints. In *Proceedings of the IEEE Electronics Components and Technology Conference (ECTC'56)*, San Diego, CA, USA, 30th May 2006, pages 250-257, 2006a.

J. Gong, C. Liu, P. P. Conway, and V. V. Silberschmidt. Modelling of Ag3Sn coarsening and its effect on creep of Sn–Ag eutectics. *Materials Science and Engineering A*, 427:60–68, 2006.

J. Gong, C. Liu, P. P. Conway, and V. V. Silberschmidt. Crystallographic structure and mechanical behaviour of SnAgCu solder interconnects under a constant loading rate. In *Proceedings of 57th Electronic Components and Technology Conference ECTC '07. Reno, NV, 29 May - 1 June 2007*, pages 677-683, 2007a.

J. Gong, C. Liu, P. P. Conway, and V. V. Silberschmidt. Micromechanical modelling of SnAgCu solder joint under cyclic loading: Effect of grain orientation. *Computational Materials Science*, 39:187–197, 2007b.

J. Gong, C. Liu, P. P. Conway, and V. V. Silberschmidt. Mesomechanical modelling of SnAgCu solder joints in flip chip. *Computational Materials Science*, 43:199–211, 2008.

M. Grediac. Principle of virtual work and identification. *Comptes Rendus de L Academie des Sciences Serie Ii*, 309:1-5, 1989.

M. Grediac, F. Auslender and F. Pierron. Applying the virtual fields method to determine the through-thickness moduli of thick composites with a nonlinear shear response. *Composites Part A - Applied Science and Manufacturing*, 32:1713-1725, 2001.

M. Grediac and F. Pierron. A T-shaped specimen for the direct characterization of orthotropic materials. *International Journal for Numerical Methods in Engineering*, 41:293-309, 1998.

M. Grediac and F. Pierron. Applying the virtual fields method to the identification of elasto-plastic constitutive parameters. *International Journal of Plasticity*, 22:602-627, 2006.

M. Grediac, F. Pierron, S. Avril and E. Toussaint, The virtual fields method for extracting constitutive parameters from full-field measurements: A review, *Strain*, 42:233-253, 2006.

M. Grediac, F. Pierron, and Y. Surrel. Novel procedure for complete in-plane composite characterization using a single T-shaped specimen. *Experimental Mechanics*, 39:142-149, 1999.

M. Grediac, F. Pierron, and A. Vautrin. The Iosipescu inplane shear test applied to composites - a new approach based on displacement field processing. *Composites Science and Technology*, 51:409-417, 1994.

M. Grediac and A. Vautrin. A new method for determination of bending rigidities of thin anisotropic plates. *Journal of Applied Mechanics-*

Transactions of the ASME, 57:964-968, 1990.

J. M. Huntley. Automated analysis of speckle interferograms. In P. K. Rastogi, editor, *Digital Speckle Pattern Interferometry and Related Techinques*, John Wiley & Sons, pages 59-139, 2001.

D. J. Inman. *Engineering Vibration*, 3rd ed., Prentice Hall; 2007.

C. Jones and C. Wykes. Electronic speckle pattern correlation interferometry. In *Holographic and Speckle Interferometry*. Cambridge, pages 165-196, 1989.

F. Jumbo, I. A. Ashcroft, A. D. Crocombe, and M. A. Wahab. The analysis of thermal residual stresses in adhesively bonded joints. In *Proceeding of the 7th European Adhesion Conference*, Freiburg, Dechema e.V., pages 532-537, 2004.

Z. L. Kahnjetter and T. C. Chu, 3-Dimensional displacement measurements using digital image correlation and photogrammic analysis. *Experimental Mechanics*, 30:10-16, 1990.

G. Kaupp and M. R. Naimi-Jamal. Nanoscratching on surfaces: the relationships between lateral force, normal force and normal displacement. *Zeitschrift für Metallkunde*. 95:297-305. 2004.

K. Kese and D. J. Rowcliffe. Nanoindentation method for measuring residual stress in brittle materials. *Journal of the American Ceramic Society*, 86: 811–816, 2003.

B. R. Lawn, A. G. Evans, and D. B. Marshall. Elastic/Plastic Indentation damage in ceramics: the median/radial crack system. *Journal of the American Ceramic Society*, 63:574-581, 1980.

B. N. Lucas and W. C. Oliver. Indentation power-law creep of high-purity indium. *Metall. Mater. Trans. A*, 30:601-610, 1999.

MSC. *Superform 2005r3 Documentation: Program Input*, 2005.

A. Maranon, P. D. Ruiz, A. D. Nurse, J. M. Huntley, L. Rivera, and G. Zhou. Identification of subsurface delaminations in composite laminates. *Composites Science and Technology*, 67:2817-2826, 2007.

A. V. Mitrofanov. *Modelling the Ultrasonically Assisted Turning of High-Strength Alloys*. PhD Thesis, Loughborough University, 2004.

W. C. Oliver and G. M. Pharr. Improved technique for determining hardness and elastic modulus using load and displacement sensing indentation experiments. *Journal of Materials Research*, 7:1564-1583, 1992.

W. C. Oliver and G. M. Pharr. Measurement of hardness and elastic modulus by instrumented indentation: Advances in understanding and refinements to methodology. *Journal of Materials Research*, 19:3-20, 2004.

W. Osten, Fringe 05. In *The 5th International Workshop on Automatic Processing of Fringe Patterns*, Springer, 2005.

W. Osten. Optical inspection of microsystems. In W. Osten, editor, *Optical Science And Engineering*, CRC/Taylor & Francis, 2007.

S. W. Park and R. A. Schapery. Methods of interconversion between linear viscoelastic material functions. Part I — A numerical method based on Prony series. *International Journal of Solids and Structures,* 36:1653-1675, 1999.

P. Pedersen and P. S. Frederiksen. Identification of orthotropic materials moduli by combined experimental numerical approach. *Measurement Science & Technology*, 10:113-118, 1992.

G. M. Pharr and A. Bolshakov. Understanding nanoindentation unloading curves. *Journal of Materials Research*, 17:2660–2671, 2002.

F. Pierron, S. Zhavoronok, and M. Grediac. Identification of the through-thickness properties of thick laminated tubes using the virtual fields method. *International Journal of Solids and Structures*, 37:4437-4453, 2000.

D. J. Popp, D. Mccormac, and G. M. Lee. Digital image registration by correlation techniques. *IEEE Transactions on Aerospace and Electronic Systems*, 9:809-809, 1973.

D. Post. *High Sensitivity Moiré.* Springer-Verlag, 1994.

D. S. Price, R. Jones, A .R. Harland, and V. V. Silberschmidt. Viscoelasticity of multi-layer textile reinforced polymer composites used in soccer balls, *Journal of Materials Science*, 43:2833–2843, 2008.

K. Ramesh. *Digital Photoelasticity: Advanced Techniques and Applications.* Springer, 2000.

P. K. Rastogi, M. Spajer, and J. Monneret. Inplane deformation measurement using holographic moiré. *Optics and Lasers in Engineering*, 2:79-103, 1981.

P. D. Ruiz, J .M. Huntley, and A. Maranon. Tilt scanning interferometry: a novel technique for mapping structure and three-dimensional displacement fields within optically scattering media. *Proceedings of the Royal Society A - Mathematical Physical and Engineering Sciences*, 462:2481-2502, 2006.

P. D. Ruiz, J. M. Huntley, and R.D. Wildman. Depth-resolved whole-field displacement measurement by wavelength-scanning electronic speckle pattern interferometry. *Applied Optics*, 44:3945-3953, 2005.

P. D. Ruiz, Y. Zhou, J. M. Huntley, and R. D. Wildman Depth-resolved whole-field displacement measurement using wavelength scanning interferometry. *Journal of Optics A: Pure and Applied Optics*, 6:679-683,2004.

J. M. Schmitt. OCT elastography: imaging microscopic deformation and strain of tissue. *Optics Express*, 3:199-211, 1998.

C. A. Sciammarella and S. K. Chawla. Holographic-Moire technique to obtain displacement components and derivatives. *Mechanics Research Communications*, 4:333-338, 1977.

R. F. Service. Battle to become the next-generation x-ray source. Science, 298:1356-1358, 2002.

V. V. Silberschmidt. Account for random microstructure in multiscale models. In

Y.W. Kwon, D.H. Allen and R. Talreja, editors. *Multiscale Modeling and Simulation of Composite Materials and Structures*. Springer, pages 1-35, 2008.

R. Sirohi. *Speckle Metrology*. (Optical Engineering, Vol. 38). Marcel Dekker Inc., 1993.

H. A. Spetzler. Holographic interferometry for measurement of small displacements. *Transactions-American Geophysical Union*, 53:511, 1972.

D. D. Steele, T. L. Chenevert, A. R. Skovoroda, and S. Y. Emelianov. Three-dimensional static displacement stimulated echo NMR elasticity imaging. *Physics in Medicine and Biology*, 45:1633–1648, 2000.

D. Stifter. Beyond biomedicine: a review of alternative applications and developments for optical coherence tomography. *Applied Physics B - Lasers and Optics*, 88:337-357, 2007.

D. Stifter, P. Burgholzer, O. Höglinger, E. Götzinger, and C. K. Hitzenberger. Polarisation-sensitive optical coherence tomography for material characterisation and strain-field mapping. *Applied Physics A - Materials Science & Processing*, 76:947-951, 2003.

D. Stifter, K. Wiesauer, R. Engelke, G. Ahrens, G. Grützner, M. Pircher, E. Götzinger, and C. K. Hitzenberger. Optical coherence tomography as a novel tool for non-destructive material characterization. *TM-Technisches Messen*, 74:51-56, 2007.

D. Stifter, K. Wiesauer, M. Wurm, E. Schlotthauer, J. Kastner, M. Pircher, E. Götzinger, and C. K. Hitzenberger. Investigation of polymer and polymer/fibre composite materials with optical coherence tomography. *Measurement Science & Technology*, 19:074011, 2008.

M. R. Strąkowski, J. Pluciński, M. Jędrzejewska-Szczerska, R. Hypszer, M. Maciejewski, and B. B. Kosmowski. Polarization sensitive optical coherence tomography for technical materials investigation. *Sensors and Actuators A - Physical*, 142:104-110, 2008.

N. W. Tschoegl. *The Phenomenological Theory of Linear Viscoelastic Behavior: An Introduction*. Springer-Verlag, 1989.

B. Vakoc, S. Yun, J. de Boer, G. Tearney, and B. Bouma. Phase-resolved optical frequency domain imaging. *Optics Express*, 13:5483-5493, 2005.

K. Wiesauer, M. Pircher, E. Götzinger, S. Bauer, R. Engelke, G. Ahrens, G. Grützner, C. K. Hitzenberger, and D. Stifter. En-face scanning optical coherence tomography with ultra-high resolution for material investigation. *Optics Express*, 13:1015-1024, 2005.

K. Wiesauer, M. Pircher, E. Götzinger, C. K. Hitzenberger, R. Engelke, G. Ahrens, G. Grützner, and D. Stifter. Transversal ultrahigh-resolution polarization-sensitive optical coherence tomography for strain mapping in materials. *Optics Express*, 14:5945-5953, 2006.

K. Wiesauer, M. Pircher, E. Götzinger, C. K. Hitzenberger, R. Oster, and D. Stifter. Investigation of glass-fibre reinforced polymers by polarisation-sensitive, ultra-high resolution optical coherence tomography: Internal structures, defects and stress. *Composites Science and Technology*, 67:3051-3058, 2007.

Y. Yasuno, Y. Sutoh, S. Makita, M. Itoh, and T.Yatagai. Polarization sensitive spectral interferometric optical coherence tomography for biological samples. *Optical Review*, 10;498-500, 2003.

J. Zhao. *Modelling Damage and Fracture Evolution in Plasma Sprayed Ceramic Coatings: Effect of Microstructure*. PhD Thesis, Loughborough University, 2005.

J. Zhao and V. V. Silberschmidt. Micromechanical analysis of effective properties of plasma sprayed alumina coatings. *Materials Science and Engineering A*, 417:287–293, 2006.

Localization and dynamic defects in lattice structures

G. S. Mishuris[a], A. B. Movchan[b], L. I. Slepyan[c]

[a] Institute of Mathematics and Physics, University of Aberystwyth, U.K.

[b] Department of Mathematical Sciences, University of Liverpool, U.K.

[c] School of Mechanical Engineering, Tel Aviv University, Israel

Abstract This review paper outlines several formulations for lattice structures where dynamic lattice Green's functions play an important role in analysis of localization near defects as well as fracture in structured media. The mechanism of dissipation discussed here is natural for lattices modelling structured media. The properties of waves generated by cracks, and the exponentially localized vibration modes near small defects are linked to spectral properties of Bloch-Floquet waves in undamaged periodic media. The results outlined in this review paper are based on the work by Slepyan (2002), Movchan & Slepyan (2007), Mishuris *et al.* (2007, 2008).

0.1 Introduction

Micro-nonuniformity of structured media results in the wave dispersion and consequently filtering and polarization phenomena observed in experiments. Another important feature of structured materials is the localization of vibrations, which may occur around defects in a periodic structure.

"Waves in lattices" is a classical topic, which is well addressed in the books by Brillouin (1953) and Maradudin *et al.* (1963). Lattice models were also used successfully in Martinsson and Movchan (2003), Jensen (2003), Cai *et al.* (2005) to design photonic/phononic crystal models based on the mass-spring interactions within periodic structures.

The range of applications for models, describing localized vibrations, extend from solid state physics problems dealing with disorder in atomic lattice structures to the design of photonic crystal fibres where localization occurs within the optical frequency range.

In photonic crystal structures, the exponentially localized defect modes occur naturally when the frequency is chosen within the stop band interval and a finite size defect is introduced into the structure. The analytical and numerical models for problems of this type are described Poulton *et al.*

(2003), where rapidly convergent series approximations for band gap Green's functions were constructed and used to characterize the defect modes. A significant progress has been made in modelling and design of photonic band gap materials, as discussed in Yablonovitch (1987, 1993) and John (1987), which was motivated by the recent advance in the technological developments of the photonic crystal fibres.

The displacement within the lattice can be written in terms Green's functions for a general type of external load. Dynamic lattice Green's functions for frequencies within the pass bands were studied by Martin (2006). For the case when the frequency has been chosen within the stop band range, the analysis of Green's functions has been published by Movchan and Slepyan (2007); this study shows the way to model exponentially localized defect modes created by a finite change of elastic stiffness or mass within the structure. It also addresses the notion of the depth of the band gap characterizing the rate of decay of exponentially localized vibration modes. Particularly new geometries of localized waveforms, referred to as the star-shaped localized solutions in lattices, are discussed in Slepyan and Ayzenberg-Stepanenko (2007).

In continuum models of cracks, singular integral equations can be used as a mathematical tool for description of the elastic displacement field; the kernel functions of such equations are represented via derivatives of Green's functions. Similar formulations are valid for the discrete systems. For certain geometries, the problems can be formulated in the form involving Fourier transforms of the unknown functions representing displacement or tractions. Functional equations of the Wiener-Hopf type for discrete systems involving semi-infinite faults (cracks) are discussed in detail in Slepyan (2002), this analysis includes both scalar and vector formulations for dynamic lattice structures. The paper by Slepyan (1981) presented the first analytical solution of a fracture model for a linear spring-like lattice in two dimensions. This work was followed by a series of publications on analytical models in dynamic spring-like lattice structures, which includes Kulakhmetova et al. (1984), Fineberg et al. (1991, 1992), Marder and Liu (1994), Marder and Gross (1995), Marder and Fineberg (1996), Sharon et al. (1996), Kessler (1999, 2000), Kessler and Levine (2001, 2003), Slepyan (2001a,b; 2005), Heizler et al. (2002), Slepyan and Ayzenberg-Stepanenko (2002, 2004, 2006).

The main difference between continuum and discrete models of fracture is related to the behavior of stress near the crack edge. While a crack in a continuous medium is characterised by the square root singularity of the stress components near its vertex, the crack growth in lattice structures is considered as a result of breakage of individual bonds within the lattice;

there is no crack edge singularity for stress in lattices. Dynamic features of the lattice structures with cracks are seen even when crack propagate quasi-statically - every bond rapture is accompanied by the discrete pulses and hence leads to the oscillations within the lattice. In turn, this may influence the direction of the propagating crack as well as the propagation speed. Models addressing these issues were studied in Slepyan (2000, 2002) and Mishuris *et al.* (2007, 2008, 2008a).

The plan of the present review paper is as follows. Section 0.2 addresses the notion of band gap Green's functions. This material is based on the recent work by Movchan and Slepyan (2007), and it concerns with the modelling of exponentially localized defect modes within lattice structures with impurities. Further, in Sections 0.3 and 0.4 we consider dynamic model problems of fracture in the lattice subjected to anti-plane shear deformation. The material of Section 0.3 is based on the results of Mishuris *et al.* (2007) and incorporates a semi-infinite crack in an inhomogeneous lattice, which includes two types of particles of different mass. The structure is chosen to model a stratified medium with the layers aligned along the horizontal direction. Such a lattice possesses a stop band within a finite range of frequencies for harmonic waves in the vertical direction. For this problem, we illustrate a general algorithm applicable for a wide range of lattices with cracks and reduce the mathematical formulation to the functional equation of the Wiener-Hopf type. We outline the possible mechanism of dissipation linked to the energy carried away from the crack by the high frequency waves. Another model of a "double fault" in the harmonic lattice, discussed in Section 0.4, introduces a configuration, which can be considered in the framework of the boundary layer approximations for a pair of two cracks in the continuous solid. This material follows the recent work by Mishuris *et al.* (2008). We also show the way to use the discrete lattice approach in order to resolve certain paradoxes, which occur naturally in some asymptotic models in continuous media.

0.2 Stop bands and localization within discrete dynamic systems

Consider a periodic linear lattice. Each particle has a unit mass and is connected to neighbouring particles by elastic springs, and it moves in the anti-plane direction according to the second Newton's law. The motion is assumed to be time-harmonic, with the radian frequency ω. Let us consider the Fourier transform $u^F(k)$ of the displacement amplitude with respect to the spatial variables; here k stands for the Bloch parameter which takes values in the reciprocal space. For the sake of simplicity, we look first at a linear, single dispersion-branch system, where homogeneous Fourier-transformed equation of motion is written in the form

$$[L(k) - \omega^2]u^F(k) = 0\,, \tag{1}$$

where $L(k) \geq 0$.

It is natural for periodic discrete systems to exhibit filtering properties, and in particular to possess stop bands of a certain frequency range. Assume, there exist stop bands $\{\omega : \omega < \omega_-\}$ and $\{\omega : \omega_+ < \omega\}$ so that for such frequencies no sinusoidal waves exist in the lattice structure. Correspondingly, the pass band frequencies are $\{\omega : \omega_- < \omega < \omega_+\}$. For the stop band values of frequency ω, equation (1) has no nontrivial real solutions k.

Let us load the system by applying a unit force to a particle located at the origin ($x = 0$) at a frequency ω within the stop band; for the discrete lattice, the corresponding solution is not singular, and we denote the displacement at the origin by $U = u(0)$, which can also be obtained via the inverse Fourier transform of

$$u^F(k) = \frac{1}{L(k) - \omega^2}\,. \tag{2}$$

We note that within the stop band range the displacement field is exponentially localized; $U > 0$ when $0 < \omega < \omega_-$, and $U < 0$ when $\omega > \omega_+$.

Next, instead of applying a force to a single particle within the lattice, we perturb the mass of this particle, and formally replace the force by the inertia of the additional mass. Using the notation M for the perturbation of the mass of the central particle and applying (1) we deduce

$$u^F(k) = \frac{M\omega^2 U}{L(k) - \omega^2}\,. \tag{3}$$

Equations (2) and (3) yield that that for frequencies within the stop band range, there exists a localized waveform coincident with the Green function, when the perturbation M of mass of the central particle is chosen in the form

$$M = \frac{1}{\Omega U}\,, \tag{4}$$

where $\Omega = \omega^2$. The notation Ω will have the same meaning in the further text below. It follows that $M > 0$ when $0 < \omega < \omega_-$, and $M < 0$ when $\omega > \omega_+$. Although for high frequencies the localized vibration modes occur when $M < 0$, the total mass of the central mass remains positive, i.e. $1 + M > 0$.

The ideas described above are applicable to a wide range of problems for continuous and lattice structures. In particular, they are still valid in analysis of localized vibration forms in multiple vibration-branch systems. Examples of such systems incorporates inhomogeneous periodic lattices, where the mass of particles or stiffness of bonds may vary within an elementary cell of periodicity.

A chain of masses on an elastic foundation

First, we illustrate the ideas outlined above for the physical example involving a uniform chain consisting of a periodic array of particles of the same mass, connected by identical elastic bonds. It is assumed that the bonds are massless, and that each particle is supported by an elastic foundation of stiffness $\varkappa$. The system is normalized in such a way that the particle mass, the bond stiffness and the cell size are defined as unity.

When a unit force of the radian frequency ω is applied to the mass located at the origin, the application of the discrete Fourier transform to the equations of motion yeilds

$$[\varkappa + 2(1 - \cos k) - \omega^2]u^F(k) = 1\,. \tag{5}$$

This lattice structure has a finite width pass band $\{\omega : \omega_- < \omega < \omega_+\}$, where

$$\omega_+ = \sqrt{4 + \varkappa} \;\text{ and }\; \omega_- = \sqrt{\varkappa}, \tag{6}$$

and the localized waveforms may exist when $0 < \omega < \omega_-$ or $\omega > \omega_+$.

When $\omega > \sqrt{4+\varkappa}$, the amplitude $u(m)$ of vibrations of the mass at $x = m$ is evaluated as follows

$$u(m) = -\frac{2^{-|m|}}{\sqrt{(\Omega-\varkappa)^2 - 4(\Omega-\varkappa)}}\Big(\sqrt{(\Omega-\varkappa)^2 - 4(\Omega-\varkappa)} \\ -(\Omega - \varkappa - 2)\Big)^{|m|}. \tag{7}$$

Indeed, the expression (7) shows that the field is exponentially localized.

If the action of the point force is replaced by the inertia associated with the perturbation M of mass at $x = 0$, then according to the algorithm described above we deduce

$$M = \frac{1}{\Omega u(0)} = -\frac{\sqrt{(\Omega-\varkappa)^2 - 4(\Omega-\varkappa)}}{\Omega} \quad \text{for } \Omega = \omega^2 > 4 + \varkappa\,. \tag{8}$$

We note that M is negative, and hence the localized oscillations exist for $\omega > \sqrt{4+\varkappa}$ if one mass within the chain is lighter than the others: $0 < 1 + M < 1$.

For sufficiently small frequencies $0 < \omega < \sqrt{\varkappa}$ the application of the unit force to the mass at $x = 0$ yields the exponentially localized solution

$$u(m) = \frac{2^{-|m|}}{\sqrt{(\varkappa - \Omega)^2 + 4(\varkappa - \Omega)}} \Big(\varkappa - \Omega + 2 \\ - \sqrt{(\varkappa - \Omega)^2 + 4(\varkappa - \Omega)}\Big)^{|m|}, \qquad (9)$$

and as before the action of the unit force can be replaced by the inertia of an additional mass placed at $x = 0$. The positive perturbation M of the mass is given by

$$M = \frac{1}{\Omega u(0)} = \frac{\sqrt{(\varkappa - \Omega)^2 + 4(\varkappa - \Omega)}}{\Omega} > 0 \quad \text{where} \quad 0 < \Omega < \varkappa. \qquad (10)$$

Indeed, in order to obtain the localized waveform by increasing the mass at the origin we require the presence of the elastic foundation. If the elastic foundation is removed, so that $\varkappa$ is replaced by zero, then the low frequency band gap disappears, and the increase of mass would not create a desired effect of localization.

Localization in a non-uniform lattice

We consider a periodic one-dimensional discrete system consisting of particles of two types linked by liner elastic springs. An elementary cell of this periodic structure contains two masses, m_1 and m_2, as shown in Fig. 1. Bloch-Floquet waves in "bi-atomic" periodic structure are very well studied in the literature (see, for example, Brillouin (1953)). With the absence of external forces, the equations of motion for the time harmonic motion of the radian frequency ω yield

$$\begin{aligned} -\omega^2 m_1 u_{1,n} &= u_{2,n} + u_{2,n-1} - 2u_{1,n}\,, \\ -\omega^2 m_2 u_{2,n} &= u_{1,n} + u_{1,n+1} - 2u_{2,n}\,, \end{aligned} \qquad (11)$$

where $u_{1,n}$ and $u_{2,n}$ are amplitudes of vibrations of masses m_1 and m_1 within the $n-$th cell of the structure. For the Bloch-Floquet wave we have

$$u_{1,m} = u_{1,0}\mathrm{e}^{ikm}\,, \quad u_{2,m} = u_{2,0}\mathrm{e}^{ikm}, \qquad (12)$$

where k is the Bloch parameter. The dispersion equation, which relates ω and k is

$$\omega^2 = \frac{m_1 + m_2}{m_1 m_2} \pm \sqrt{\left(\frac{m_1 + m_2}{m_1 m_2}\right)^2 - \frac{4\sin^2 k/2}{m_1 m_2}}\,. \qquad (13)$$

If the masses are equal then the problem is reduced to the case of the previous section when $\varkappa = 0$. Assume that the mass m_1 is greater than m_2. Then the system has two stop bands, i.e. the intervals of frequencies where the dispersion equation (13) does not have real roots k. It is convenient to introduce the notations:

$$\omega_- = \sqrt{\frac{2}{m_1}}, \quad \omega_+ = \sqrt{\frac{2}{m_2}}. \tag{14}$$

Then according to the above dispersion relation the frequency band gaps are

$$\Pi_1 = \{\omega : \omega_- < \omega < \omega_+\}, \tag{15}$$

and

$$\Pi_2 = \{\omega : \omega > \sqrt{\frac{2}{m_1} + \frac{2}{m_2}}\}. \tag{16}$$

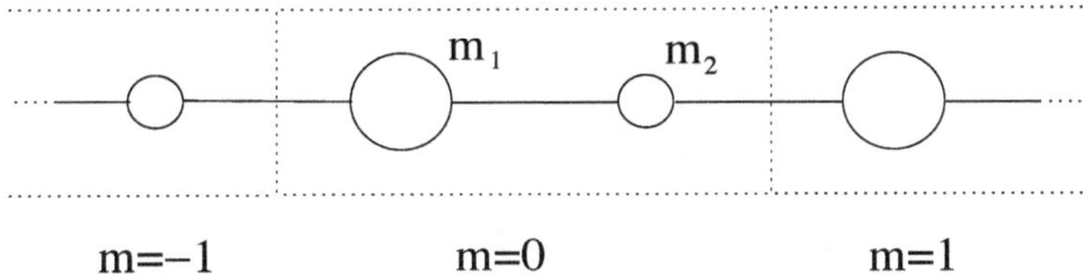

Figure 1. A bi-atomic chain.

Band gap Green's function for inhomogeneous lattice

Now, for frequencies with the band gaps, we consider the Green's matrix function. The force can be applied to any of the two particles within the elementary cell. Let P_1 be the value of the force applied to the mass m_1 and P_2 be the force applied to the mass m_2; also as before we shall use the notation $\Omega = \omega^2$. The equations of motion yield

$$\begin{aligned} -m_1\Omega u_{1,m} &= u_{2,m} + u_{2,m-1} - 2u_{1,m} + P_1\delta_{0,m}, \\ -m_2\Omega u_{2,m} &= u_{1,m} + u_{1,m+1} - 2u_{2,m} + P_2\delta_{0,m}. \end{aligned} \tag{17}$$

The discrete Fourier transform leads to the solution

$$u_1^F = \frac{1}{Q}[-P_1(m_2\Omega - 2) + P_2(1 + e^{ik})],$$
$$u_2^F = \frac{1}{Q}[-P_2(m_1\Omega - 2) + P_1(1 + e^{-ik})], \tag{18}$$

where

$$Q = (m_1\Omega - 2)(m_2\Omega - 2) - 2(1 + \cos k), \tag{19}$$

where $Q < 0$ for the finite band gap Π_1 and $Q > 0$ for the high-frequency band gap Π_2.

Applying the inverse Fourier transform we obtain the displacement amplitudes

$$u_{1,m}(P_1, P_2) = -\frac{1}{\sqrt{Q_0^2 - 4Q_0}}\{[-P_1(m_2\Omega - 2) + P_2]\lambda^{|m|} + P_2\lambda^{|m-1|}\},$$
$$u_{2,m}(P_1, P_2) = -\frac{1}{\sqrt{Q_0^2 - 4Q_0}}\{[-P_2(m_1\Omega - 2) + P_1]\lambda^{|m|} + P_1\lambda^{|m+1|}\}, \tag{20}$$

where

$$\lambda = \frac{1}{2}\left(\sqrt{Q_0^2 - 4Q_0} + Q_0 - 2\right) \quad \text{and}$$
$$Q_0 = (m_1\Omega - 2)(m_2\Omega - 2) < 0 \quad \text{for} \quad \frac{2}{m_1} < \Omega < \frac{2}{m_2}. \tag{21}$$

We note that $-1 < \lambda < 0$, and hence when the frequency is chosen within the finite stop band Π_1 the solution (28) is exponentially localised. The Green's matrix function is then given by

$$\mathbf{G}_m = \begin{pmatrix} u_{1,m}(1,0) & u_{1,m}(0,1) \\ u_{2,m}(1,0) & u_{2,m}(0,1) \end{pmatrix}.$$

In particular, for $m = 0$, we obtain the amplitudes of the particle displacements in the central cell

$$u_{1,0} = \frac{P_1(m_2\Omega - 2) - (1/2)P_2\left(Q_0 + \sqrt{Q_0^2 - 4Q_0}\right)}{\sqrt{Q_0^2 - 4Q_0}},$$
$$u_{2,0} = \frac{P_2(m_1\Omega - 2) - (1/2)P_1\left(Q_0 + \sqrt{Q_0^2 - 4Q_0}\right)}{\sqrt{Q_0^2 - 4Q_0}}. \tag{22}$$

Localized vibration modes within the finite band gap

Next, instead of applying the forces to the particles of the central cell ($m = 0$), we introduce perturbations M_1 and M_2 of masses for the first and second particle, respectively. In this case, two particles contained in the central cell will have masses $m_1 + M_1$ and $m_2 + M_2$; the masses of other particles within the chain remain unchanged.

In this case, we consider free vibrations of the new lattice structure, and if the localized waveform exists then the results outlined in the previous section are applicable, with the formal replacement $P_1 = M_1 \Omega u_{1,0}$, $P_2 = M_2 \Omega u_{2,0}$ in formulae (28). For the radian frequency ω chosen within the finite width stop band Π_1 we have $\frac{2}{m_1} < \Omega < \frac{2}{m_2}$ and write the following homogeneous system of linear algebraic equations

$$u_1 = \frac{M_1 \Omega u_1 (m_2 \Omega - 2) - (1/2) M_2 \Omega u_2 \left(Q_0 + \sqrt{Q_0^2 - 4Q_0} \right)}{\sqrt{Q_0^2 - 4Q_0}},$$

$$u_2 = \frac{M_2 \Omega u_2 (m_1 \Omega - 2) - (1/2) M_1 \Omega u_1 \left(Q_0 + \sqrt{Q_0^2 - 4Q_0} \right)}{\sqrt{Q_0^2 - 4Q_0}}. \quad (23)$$

Note that for the sake of simplicity $u_{1,0}, u_{2,0}$ have been replaced by u_1, u_2. The system (23) has non-trivial solutions if and only if

$$\left(\sqrt{Q_0^2 - 4Q_0} - M_1 \Omega (m_2 \Omega - 2) \right) \left(\sqrt{Q_0^2 - 4Q_0} - M_2 \Omega (m_1 \Omega - 2) \right) - \frac{M_1 M_2 \Omega^2}{4} \left(\sqrt{Q_0^2 - 4Q_0} + Q_0 \right)^2 = 0. \quad (24)$$

which gives a constraint on the choice of "additional masses" M_1 and M_2.

The frequency of the localized waveform depends on perturbations M_1 and M_2 of masses within the central cell, as shown in two examples below.

Design of localized waveforms via perturbation of mass

Here we show that the localized defect mode can be created via a small variation of the masses in the central cell.

Example 1. First, consider the case when the second mass remains intact ($M_2 = 0$), whereas the perturbation M_1 is chosen to be consistent with the localized vibration of the radian frequency $\omega \in \Pi_1$. For the sake of convenience, we introduce the normalization of the lattice, so that the total mass of the unperturbed elementary cell is $m_1 + m_2 = 2$. Also, let us define the contrast parameter $r = m_1/m_2 > 1$. Then the required perturbation M_1 is determined by

$$\begin{aligned} M_1 &= -\frac{1}{\omega^2}\sqrt{\frac{m_1\omega^2 - 2}{2 - m_2\omega^2}}\sqrt{4 - Q_0} = -\frac{2}{\Omega(1+r)}\sqrt{\frac{r\Omega - 1 - r}{1 + r - \Omega}} \\ &\quad \times\sqrt{(1+r)^2 + (r\Omega - 1 - r)(1 + r - \Omega)}\ , \qquad (25) \end{aligned}$$

where

$$Q_0 = 4\left(\frac{\Omega}{1+r} - 1\right)\left(\frac{r\Omega}{1+r} - 1\right). \qquad (26)$$

For a small change of the first mass, a localized mode appears near the lower edge of the band gap, i.e. $\Omega \to 2m_1^{-1} + 0$, when $M_2 = 0$ and $M_1 \sim -C\sqrt{m_1\Omega - 2}$, with C being a positive constant. In Fig. 2, we present the graphs of frequencies of the localized modes as functions of the mass ratio in the unperturbed bi-atomic lattice for the case when $M_2 = 0$ and M_1 is negative. This includes the curves corresponding to $M_1 = -0.5,\ -0.7,\ -1$, and the limit curve $M_1 = -2r/(1+r)$, where $m_1 + M_1 = 0$. This illustration suggests that by decreasing the first mass, while $M_1 + m_1 > 0$ and the other mass is kept unchanged, we can design a localized defect mode at the frequency which lies within the region between the limit curve and the lower edge of the band gap, where $\Omega = 2/m_1 = 1 + 1/r$. The localized mode near the upper edge of the band gap cannot be created by a small variation of m_1 while $M_2 = 0$.

Example 2. Now consider the situation when the mass of the first particle remains unchanged, i.e. $M_1 = 0$, but the second mass is perturbed according to the formula

$$\begin{aligned} M_2 &= \frac{1}{\Omega(1+r)}\sqrt{\frac{2 - m_2\Omega}{m_1\Omega - 2}}\sqrt{4 - Q_0} \\ &= \frac{2}{\Omega(1+r)}\sqrt{\frac{1 + r - \Omega}{r\Omega - 1 - r}}\sqrt{(1+r)^2 + (r\Omega - 1 - r)(1 + r - \Omega)} \qquad (27) \end{aligned}$$

This gives another rule for generating an exponentially localized waveform for a lattice of the given contrast r, at a frequency within the stop band Π_1. We note that the localized mode near the lower edge of the band gap cannot be created by a small variation of m_2 with $M_1 = 0$.

On the contrary, to obtain a localized vibration at the frequency close to the upper edge of the band gap, i.e. $\Omega \to 2m_2^{-1} - 0$, it is sufficient

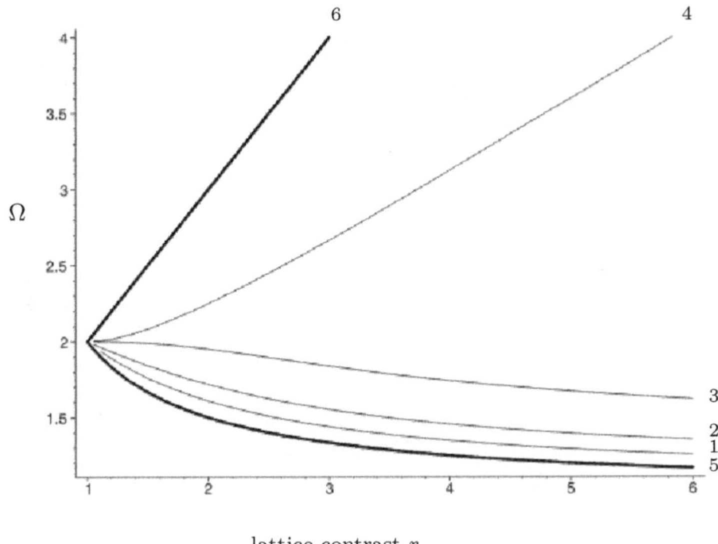

Figure 2. The graphs of Ω versus the lattice contrast r for $M_2 = 0$: (1) $M_1 = -0.5$; (2) $M_1 = -0.7$; (3) $M_1 = -1$; (4) the limit curve $M_1 = -2r/(1+r)$; (5) the lower band gap boundary; (6) the upper band gap boundary.

to increase the smaller mass m_2, so that $M_2 \sim C\ \sqrt{2 - m_2\Omega}$, $C > 0$, while $M_1 = 0$. In Fig. 3, we show the frequencies of the localized modes as functions of the mass ratio $r = m_1/m_2$ for the case when $M_1 = 0$ and M_2 is positive. The diagram incorporates the curves corresponding to $M_2 = 0.5,\ 1,\ 2,\ 3$. The open band gap region can be covered via increase of the mass m_2 while $M_1 = 0$. However, in order to reach the lower bound of the band gap one has to take the limit as $m_2 \to +\infty$.

Simultaneous finite variations of both masses, m_1 and m_2 allow to create a localized mode over all the frequency range within the finite band gap Π_1.

High frequency localized modes

The analysis produced for frequencies within the finite width band gap Π_1 can be extended to the case of high frequencies from the interval Π_2, when $\Omega > \Omega_{max} = 2/m_1 + 2/m_2 = 2 + r + 1/r$.

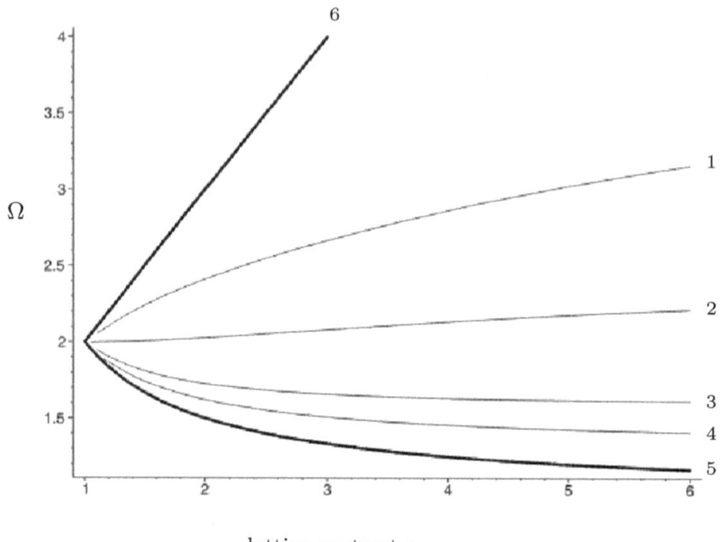

Figure 3. The graphs of Ω versus the lattice contrast r for $M_1 = 0$: (1) $M_2 = 0.5$; (2) $M_2 = 1$; (3) $M_2 = 2$; (4) $M_2 = 3$; (5) the lower band gap boundary; (6) the upper band gap boundary.

For the forced system, it follow from equations (19) that the amplitudes of vibrations are given by

$$\begin{aligned} u_{1,m} &= \frac{1}{\sqrt{Q_0^2 - 4Q_0}}\{[-P_1(m_2\Omega - 2) + P_2]\lambda^{|m|} + P_2\lambda^{|m-1|}\}, \\ u_{2,m} &= \frac{1}{\sqrt{Q_0^2 - 4Q_0}}\{[-P_2(m_1\Omega - 2) + P_1]\lambda^{|m|} + P_1\lambda^{|m+1|}\}, \end{aligned} \tag{28}$$

where

$$\begin{aligned} 0 < \lambda = \frac{1}{2}\left(Q_0 - 2 - \sqrt{Q_0^2 - 4Q_0)}\right) < 1 \\ \text{and} \qquad Q_0 = (m_1\Omega - 2)(m_2\Omega - 2) > 0. \end{aligned} \tag{29}$$

Hence the displacement amplitudes (28) decay exponentially as $|m| \to \infty$.

The amplitudes in the central cell are

$$u_{1,0} = \frac{-P_1(m_2\Omega - 2) + (1/2)P_2\left(Q_0 - \sqrt{Q_0^2 - 4Q_0}\right)}{\sqrt{Q_0^2 - 4Q_0}},$$

$$u_{2,0} = \frac{-P_2(m_1\Omega - 2) + (1/2)P_1\left(Q_0 - \sqrt{Q_0^2 - 4Q_0}\right)}{\sqrt{Q_0^2 - 4Q_0}}. \tag{30}$$

To model a defect mode we formally replace P_1, P_2 according to the rule $P_1 = M_1\Omega u_{1,0}$, $P_2 = M_2\Omega u_{2,0}$, and hence derive a system of two linear equations with respect to $u_{1,0}$ and $u_{2,0}$, which has a non-trivial solution if and only if

$$\left(\sqrt{Q_0^2 - 4Q_0} + M_1\Omega(m_2\Omega - 2)\right)\left(\sqrt{Q_0^2 - 4Q_0} + M_2\Omega(m_1\Omega - 2)\right) - \frac{M_1 M_2 \omega^4}{4}\left(Q_0 - \sqrt{Q_0^2 - 4Q_0}\right)^2 = 0 \;, \tag{31}$$

where

$$\Omega > \frac{2}{m_1} + \frac{2}{m_2} = \frac{(1+r)^2}{r}. \tag{32}$$

It follows that a high-frequency localized waveform can be created by reducing one of the masses within the central cell. For any frequency within the interval Π_2, the appropriate change of mass, generating the defect mode, can be determined as follows.

If the first mass m_1 is replaced by $m_1 + M_1 > 0$, while the second mass is unchanged ($M_2 = 0$), then for a given $\omega \in \Pi_+$ the perturbation $M_1 < 0$ is defined by

$$M_1 = -\frac{1}{\omega^2}\sqrt{\frac{m_1\omega^2 - 2}{m_2\omega^2 - 2}}\sqrt{Q_0 - 4} > -m_1. \tag{33}$$

On the other hand, if the first mass remains unchanged ($M_1 = 0$), then to create a localized defect mode one would need to perturb the second mass by the amount $M_2 < 0$, where

$$M_2 = -\frac{1}{\Omega}\sqrt{\frac{m_2\Omega - 2}{m_1\Omega - 2}}\sqrt{Q_0 - 4} > -m_2 \;. \tag{34}$$

0.3 Dynamic fault in a two-dimensional inhomogeneous lattice structure

The ideas of the previous sections will be extended to the case of an inhomogeneous lattice which occupies an infinite plane. Non-uniformity within the lattice is introduced in the same way as before, i.e. by assigning different masses to the joints of the lattice. The problem corresponds to the anti-plane deformation, and the elementary cell of the periodic lattice has a rectangular shape, as shown in Fig. 4a. In this diagram, the particles represented by black (or white) discs are assumed to have the mass m_1 (or m_2). The rows of joints of the same mass are aligned with the horizontal axis, and with our choice of distribution of mass, an elementary cell of the doubly periodic structure contains three particles, with two particles of mass m_1 and one particle of mass m_2.

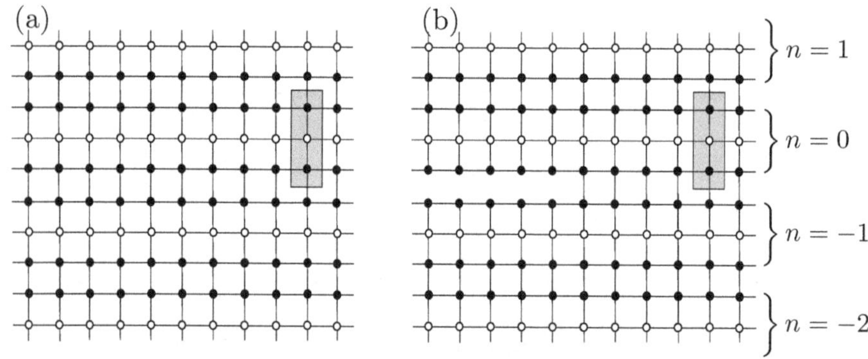

Figure 4. Nonuniform lattice structures: (a) Undamaged lattice, (b) Lattice with a crack. The elementary cell is shown as a shaded rectangle. The horizontal and vertical coordinates of the cell are denoted by m and n, respectively.

Similar to the previous text, we introduce the normalization in such a way that the stiffness of the bonds connecting neighbouring particles, the lattice spacing between neighbouring particles, and the averaged mass density within the elementary cell are equal to unity. This implies that $(2m_1 + m_2)/3 = 1$, and that the low-frequency wave speed is equal to $c = 1$. The notation $r = m_1/m_2$ is used for the contrast parameter of the lattice.

The equations of motion are written for three particles within the elementary cell of the periodic structure, and the displacement of a node is denoted by $u_{j,m,n}$, where (m, n), $m, n = 0, \pm 1, \pm 2, ...$, is the multi-index characterising the position of the cell $me^{(1)} + 3ne^{(2)}$, and the remaining index $j = 0, 1, 2$, characterises the position of the particle within the cell:

$$\begin{aligned}
\frac{3r}{2r+1}\ddot{u}_{0,m,n} &= u_{0,m-1,n} + u_{0,m+1,n} + u_{1,m,n} + u_{2,m,n-1} - 4u_{0,m,n}\,, \\
\frac{3}{2r+1}\ddot{u}_{1,m,n} &= u_{1,m-1,n} + u_{1,m+1,n} + u_{2,m,n} + u_{0,m,n} - 4u_{1,m,n}\,, \qquad (35) \\
\frac{3r}{2r+1}\ddot{u}_{2,m,n} &= u_{2,m-1,n} + u_{2,m+1,n} + u_{0,m,n+1} + u_{1,m,n} - 4u_{2,m,n}.
\end{aligned}$$

Note that when a fault is introduced into the lattice the equations change for particles placed on the boundary of the fault. The fault is defined as an array of broken bonds, and we call it a crack.

We assume that a semi-infinite crack, $m < vt, t > -\infty$, shown in Fig. 4b, propagates along a line between the layers $(0, m, 0)$ and $(2, m, -1)$, with a constant speed $v > 0$. In such a "steady state", it is assumed that the displacement $u_{j,m,n}(t)$ depends only on the variables j, $\eta = m - vt$ and n. The crack propagation is accompanied by the bond breakage at $\eta = 0$, and it is convenient to describe the displacement field by functions $u_{j,n}(\eta)$. Assuming the mode III symmetry we can state

$$\begin{aligned}
&u_{0,n}(\eta) = -u_{2,-n-1}(\eta)\,, \quad u_{1,n}(\eta) = -u_{1,-n-1}(\eta)\,, \\
&u_{2,n}(\eta) = -u_{0,-n-1}(\eta)\,. \qquad (36)
\end{aligned}$$

On the upper and lower crack faces, the displacements are $u_{0,0}(\eta)$ and $u_{2,-1}(\eta)$, respectively. The symmetry of the problem implies that $u_{2,-1}(\eta) = -u_{0,0}(\eta)$. It is assumed that the crack faces are traction free, so that if σ denotes the traction on the boundary of the upper half-plane, then for $\eta < 0$ we have

$$\sigma(\eta) = 0\,, \qquad (37)$$

whereas for $\eta > 0$

$$\sigma(\eta) = 2u_{0,0}(\eta)\,. \qquad (38)$$

Similar to Section 0.2, we first consider Bloch-Floquet waves in the infinite two-dimensional periodic lattice. Furthermore, we reduce the problem for a semi-infinite crack in the lattice to a functional equation of the Wiener-Hopf type.

Bloch-Floquet waves in the doubly periodic lattice

First we consider the undamaged discrete structure. Bloch-Floquet waves in the lattice, whose elementary cell consists of three particles, are sought in the form:

$$u_{j,m,n} = U_j e^{i(\omega t - k_x m - 3k_y n)}, \tag{39}$$

where U_j, $j = 0, 1, 2$, are the amplitudes, ω is the radian frequency, and (k_x, k_y) is the Bloch vector in the reciprocal plane possessing the periodic structure with the elementary cell $(-\pi, \pi] \times (-\pi/3, \pi/3]$.

The system (35) yields

$$\begin{aligned} S_2 U_0 - U_1 - U_2 e^{3ik_y} &= 0, \\ S_1 U_1 - U_2 - U_0 &= 0, \\ S_2 U_2 - U_0 e^{-3ik_y} - U_1 &= 0, \end{aligned} \tag{40}$$

where

$$S_1 = 4 - 2\cos k_x - \frac{3}{2r+1}\omega^2, \quad S_2 = 4 - 2\cos k_x - \frac{3r}{2r+1}\omega^2. \tag{41}$$

This system of linear algebraic equations with respect to U_0, U_1, U_2 has a nontrivial solution if and only if

$$S_1(S_2^2 - 1) - 2S_2 - 2\cos(3k_y) = 0, \tag{42}$$

which is a cubic algebraic equation with respect to $\Omega = \omega^2$.

The dispersion surfaces $\omega = \omega(k_x, k_y)$ representing the solutions of the dispersion equation (42) are shown in Fig. 5, and for a non-homogeneous lattice, when $r \neq 1$, we deal with a three-branch system.

For long waves, as $k_x, k_y \to 0$, the corresponding dispersion surface has a conical shape, as expected for the homogenized medium

$$\omega^2 \sim k_x^2 + k_y^2. \tag{43}$$

The Wiener-Hopf equation for the semi-infinite fault

When the moving crack is introduced, we break the periodicity of the lattice. Nevertherless, the information about the Bloch-Floquet waves outlined in the previous section remains useful, as we are going to show in this section. As mentioned above, the dynamic field around the crack is described in terms of the displacement function $u_{j,n}(\eta)$, where $\eta = m - vt$ and v is the crack velocity. We formally apply the Fourier transform with respect to η considered as a continuous variable. Using the notation $u_{j,n}^F(k)$ for the Fourier transform of $u_{j,n}(\eta)$, we write the Fourier tranformed equations of

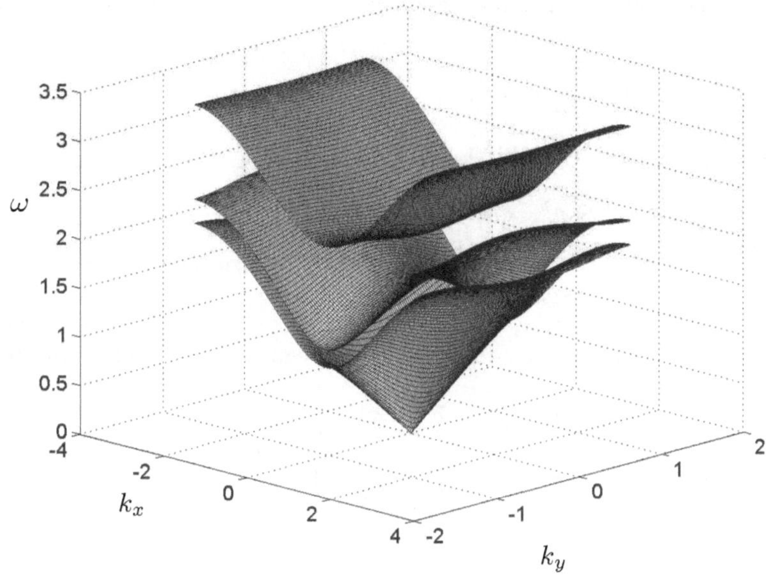

Figure 5. The three-branch dispersion diagram; the contrast parameter is $r = 2$.

motion for a lattice above the crack ($n > 0$) in the form

$$\begin{aligned} \mathcal{S}_2 u_{0,n}^F - u_{1,n}^F - u_{2,n-1}^F &= 0\,, \\ \mathcal{S}_1 u_{1,n}^F - u_{0,n}^F - u_{2,n}^F &= 0\,, \\ \mathcal{S}_2 u_{2,n}^F - u_{0,n+1}^F - u_{1,n}^F &= 0\,, \end{aligned} \tag{44}$$

with

$$\begin{aligned} \mathcal{S}_1(k) &= 4 - 2\cos k + \frac{3}{2r+1}(0 + ikv)^2\,, \\ \mathcal{S}_2(k) &= 4 - 2\cos k + \frac{3r}{2r+1}(0 + ikv)^2\,, \end{aligned} \tag{45}$$

where

$$(0 + ikv) = \lim_{\varepsilon \to +0} (\varepsilon + ikv). \tag{46}$$

It is worth noting that the expressions (45) are similar to those in (41) with the formal replacement of $-\omega^2$ by $(0 + ikv)^2$.

For the upper half-plane, when $n > 0$, we represent the solution of (44) in the form

$$u_{j,n}^F(k) = u_j^F(k)\lambda^n(k)\,, \quad u_j^F = u_{j,0}^F\,, \quad |\lambda| \le 1\,, \tag{47}$$

and then equations (44) yield

$$\begin{aligned} \mathcal{S}_2 u_0^F - u_1^F - u_2^F/\lambda &= 0\,, \\ \mathcal{S}_1 u_1^F - \; u_0^F - u_2^F &= 0\,, \\ \mathcal{S}_2 u_2^F - \lambda u_0^F - u_1^F &= 0\,. \end{aligned} \tag{48}$$

The linear algebraic system (48) has a nontrivial solution if and only if

$$\det \begin{pmatrix} \mathcal{S}_2 & -1 & -\lambda^{-1} \\ -1 & \mathcal{S}_1 & -1 \\ -\lambda & -1 & \mathcal{S}_2 \end{pmatrix} = 0. \tag{49}$$

It follows that

$$\lambda = P - \sqrt{P^2 - 1}\,, \qquad P = \frac{1}{2}\left(\mathcal{S}_1\mathcal{S}_2^2 - 2\mathcal{S}_2 - \mathcal{S}_1\right)\,, \tag{50}$$

and the equations (42) and (49) coincide if we formally take

$$\omega = kv \quad \text{and} \quad k_y = \frac{\mathrm{i}}{3}\ln\lambda(k), \tag{51}$$

where $k = k_x$. If we look at the surfaces in the three-dimensional space with the axes k_x, k_y, ω, then the intersection of the plane $\omega = k_x v$ with the dispersion surfaces in Fig. 5 and the surfaces defined by the last equality in (51) define the waves which can be excited by the propagating crack. Indeed, if $|\lambda| = 1$ or $|\lambda| < 1$, then the corresponding solution represents sinusoidal or exponentially decreasing waves, respectively.

For the crack problem, it is essential to determine the displacement $u_{0,0}(\eta)$ on the crack faces. In terms of the Fourier transforms, we use the representation

$$u_{0,0}^F(k) = u_+(k) + u_-(k)\,, \quad u_\pm(k) = \left[u_{0,0}(\eta)H(\pm\eta)\right]^F\,, \tag{52}$$

where H is the Heaviside function. Since the crack faces are traction free, we have

$$\sigma(\eta) = 0\,, \quad \sigma_- = 0 \;\text{ for }\; \eta < 0\,, \tag{53}$$

and

$$\sigma(\eta) = 2u_{0,0}(\eta)\,, \quad \sigma_+ = 2u_+ \text{ for } \eta > 0\,. \tag{54}$$

Then using the relation

$$(\mathcal{S}_1\mathcal{S}_2 - 1)u_1^F = (\mathcal{S}_2 + \eta)u_0^F \tag{55}$$

we derive the functional equation of the Wiener-Hopf type, which relates the functions $u_+(k)$ and $u_-(k)$

$$\mathcal{Q}_1(k)u_+(k) + \mathcal{Q}_2(k)u_-(k) = 0\,,$$
$$\mathcal{Q}_1(k) = \mathcal{S}_2 + 1 - \frac{\mathcal{S}_2 + \lambda}{\mathcal{S}_1\mathcal{S}_2 - 1}\,, \quad \mathcal{Q}_2(k) = \mathcal{S}_2 - 1 - \frac{\mathcal{S}_2 + \lambda}{\mathcal{S}_1\mathcal{S}_2 - 1}\,, \tag{56}$$

where $\mathcal{S}_{1,2}$ and λ are the same as in (45) and (50).

We now define the function

$$L(k) = \mathcal{Q}_1/\mathcal{Q}_2 \tag{57}$$

as the kernel of the Wiener-Hopf equation written in the form

$$L(k)u_+(k) + u_-(k) = 0\,. \tag{58}$$

For the lattice of a finite contrast, the index of $L(k)$ is equal to zero (see Slepyan (2002), p. 451)), and $L(\pm\infty) = 1$. Following the standard factorization procedure we deduce

$$L(k) = L_+(k)L_-(k)\,,$$
$$L_\pm(k) = \exp\left[\pm\frac{1}{2\pi \mathrm{i}}\int_{-\infty}^{\infty}\frac{\ln L(\xi)}{\xi - k}\,\mathrm{d}\xi\right]\,, \tag{59}$$

where L_+ is analytic for $\Im k > 0$, and L_- is analytic for $\Im k < 0$.

The factorized equation for u_+ and u_- takes the form

$$L_+(k)u_+(k) + L_-^{-1}u_-(k) = 0\,. \tag{60}$$

Following Slepyan (2002), Chapter 14, and Mishuris *et al.* (2007), we modify this equation by allowing for a contribution from the remote external load, which will be represented by the δ-function type term in the right-hand side:

$$L_+(k)u_+(k) + L_-^{-1}u_-(k) = C\left[(0 + \mathrm{i}k)^{-1} + (0 - \mathrm{i}k)^{-1}\right]\,, \tag{61}$$

where C is a constant. The solution, which satisfies the regularity condition including the boundedness of the displacements at the crack edge, has the form

$$u_+(k) = \frac{C}{(0 - ik)L_+(k)}, \quad u_-(k) = \frac{CL_-(k)}{0 + ik}. \tag{62}$$

The behaviour of the solution near $k = 0$ defines the long wave carrying energy to the propagating crack from $-\infty$. On the other hand, the real singular points of $L_\pm(k)$ and zeros of $L_+(k)$ correspond to the high-frequency waves excited by the propagating crack.

Wave tunneling along the crack

Similar to Section 0.2, where we have seen exponentially localized solutions for dynamic problems in nonhomogeneous lattice structures, one can ask if similar effects may appear for two-dimensional lattice structures with defects. While for the model of Section 0.2 the rate of exponential decay was determined by the band gap Green's function, for the present problem involving the moving crack in the lattice the wave tunneling is controlled by the kernel function L of the Wiener-Hopf equation (58).

Here we highlight a class of solutions, which correspond to waves tunneled along the crack. These solutions are also used in the dissipation mechanism for lattice structures, as outlined in the text below. The attention is given to the roots of $\mathcal{Q}_1(k)$ or $\mathcal{Q}_2(k)$ from (56), which correspond to singular points of $u_+(k)$ or $u_-(k)$. Three set of roots of the equations

$$\mathcal{Q}_1(k) = 0 \quad \text{and} \quad \mathcal{Q}_2(k) = 0 \tag{63}$$

for a non-homogeneous lattice ($r \neq 1$) are determined and analyzed in Mishuris *et al.* (2007) for

$$\lambda = \pm 1 .$$

It has also been shown that, for these waves, which can be referred to as "dissipative waves", there is no averaged energy flux in y-direction, i.e. the y-component of the group velocity of these waves is equal to zero. Thus, these waves carry the energy parallel to the crack. Nevetherless, since oscillations percolate in the y-direction, the steady-state wave field is not localized near the crack.

Energy dissipation in the lattice containing a moving crack

Slepyan (2002) introduced the notion of the local energy release rate G_0 for the lattice and the global energy release rate G be for the corresponding homogenized medium, in such a way that G_0 is the strain energy accumulated in the bond before it breaks, whereas G is the energy release rate corresponding to the long-wave mode of the exact lattice solution. In this case, the difference, $G - G_0$, represents the energy of high-frequency waves radiated by the propagating crack. This quantity can be interpreted (see Slepyan (2002) and Mishuris *et al.* (2007)) as the dissipation energy release rate, and it has been shown that

$$\frac{G_0}{G} = \mathcal{R}^2(v), \tag{64}$$

with

$$\mathcal{R}(v) = \exp\left[\frac{1}{\pi}\int_0^\infty \frac{\operatorname{Arg} L(k)}{k}\,\mathrm{d}k\right], \tag{65}$$

where $\operatorname{Arg} L(k)$ depends on the lattice structure, and, in particular, it depends on the lattice contrast parameter r.

The limiting elongation of the bond is equal to $2u(\eta)$ at $\eta = 0$, and

$$2u(0) = 2\lim_{k\to \mathrm{i}\infty}(-\mathrm{i}k)u_+(k) = 2C\ , \tag{66}$$

and hence, in our solution, the limiting energy of the linearly elastic bond is given by

$$G_0 = 2C^2\,. \tag{67}$$

At the same time, the long-wave behaviour ($k \to 0$) of σ_+ and u_- are defined by the same formulas as for the homogeneous lattice (compare with Slepyan (2002), Section 11.5.2)

$$\sigma_+(k) \sim \frac{\sqrt{2}C(1-v^2)^{1/4}}{\mathcal{R}(v)\sqrt{0-\mathrm{i}k}}\,,$$
$$u_-(k) \sim \frac{\sqrt{2}C}{(1-v^2)^{1/4}\mathcal{R}(v)(0+\mathrm{i}k)^{3/2}}, \tag{68}$$

with

$$\mathcal{R}(v) = \exp\left[\frac{1}{\pi}\int_0^\infty \frac{\operatorname{Arg} L(k)}{k}\,\mathrm{d}k\right], \tag{69}$$

where $\operatorname{Arg} L(k)$ depends on the lattice structure.

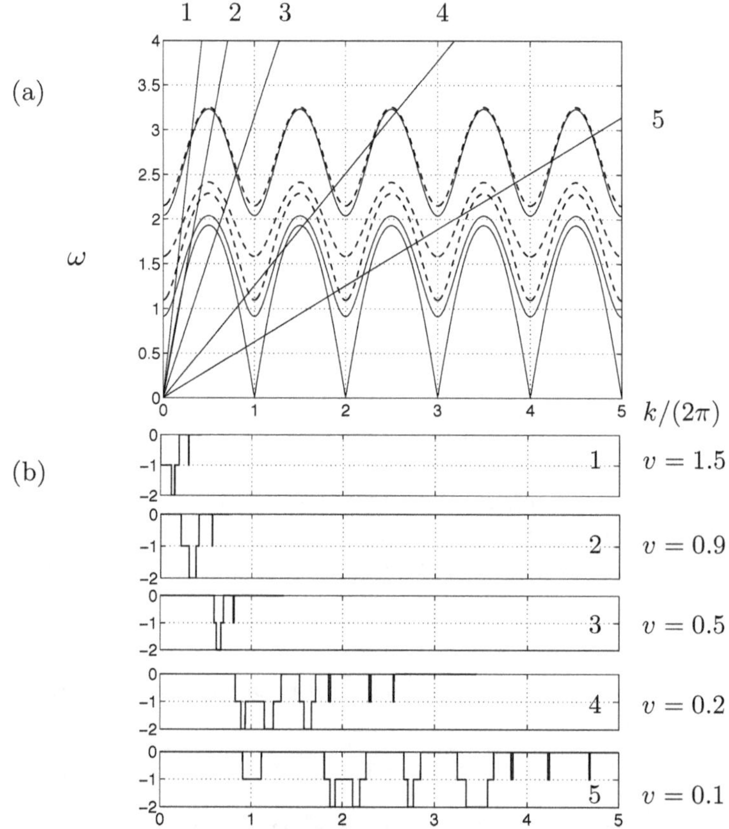

Figure 6. (a) Dispersion diagrams together with the rays $\omega = kv$. (b) The corresponding normalised argument $2\pi^{-1}\mathrm{Arg}L(k)$ for $r = 2$. The ray numbers, $j = 1, 2, ..., 5$, on the dispersion diagram (a) are repeated on the diagrams (b). Here and in the following figures, the dashed dispersion curves correspond to zeros of $\mathcal{Q}_1(k)$, whereas the solid curves correspond to zeros of $\mathcal{Q}_2(k)$.

The diagrams including the dispersion curves, intersecting with the rays $\omega = kv$ for certain fixed values of the crack speed v, as well as the corresponding diagrams for $\mathrm{Arg}\, L(k)$ are presented in Figs. 6a and 6b, respectively.

Results of calculations of the energy ratio, $\mathcal{R}^2(v)$ for a number of values of the lattice contrast μ, are presented in Fig. 7. Here we use the contrast variable $\mu = 1/r$. An interesting result obtained in Mishuris *et al.* (2007) is a considerable *energy ratio drop-down* which arises in the energy ratio diagram for a certain region of the lattice contrast. It mainly arises when $\mu > 1$, and the corresponding region on the v-axis moves to the right as μ increases (see the diagrams in Fig. 7 for $\mu = 2, 5, 10, 20$). With the further increase of μ, this *drop-down* approaches the critical crack speed where the energy ratio equals zero.

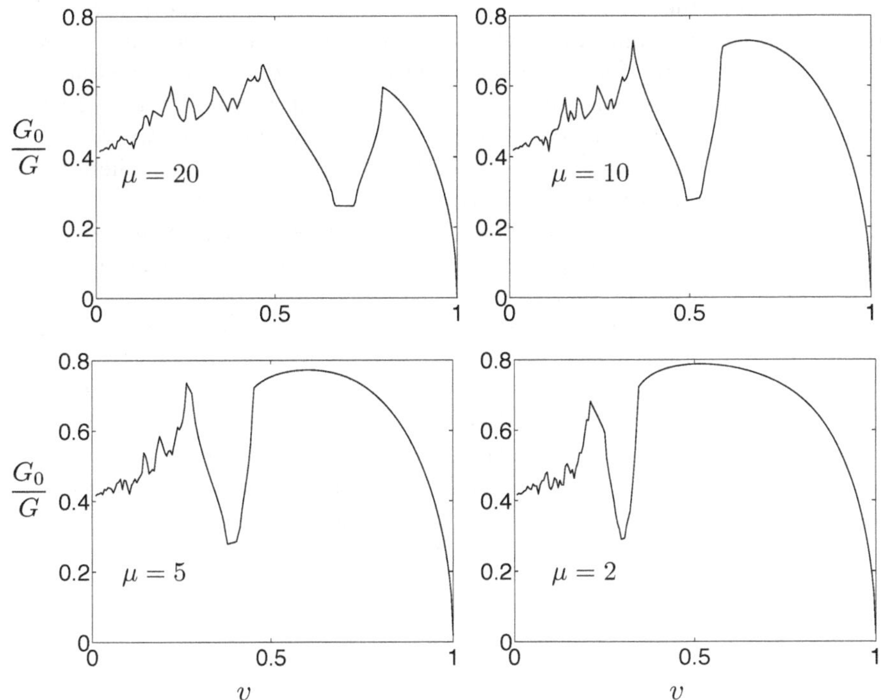

Figure 7. The energy ratio as a function of the crack speed v for different values of the lattice contrast parameter $\mu = 1/r$.

0.4 Lattice models versus paradoxes of continuous formulations

On some occasions, lattice models can serve as boundary layer formulations for problems in a medium with a micro-structure. If the boundary layer is not taken into account, the continuous solutions may exhibit unrealistic behaviour of different types. However, with the lattice solutions in place, the "paradox" gets resolved.

In this section we consider one of such examples involving the propagation of a symmetric double-crack fault shown in Fig. 8. Here, the chain of masses connected by massless bonds, is removed by a remote force, applied in the out-of-plane mode, from the lattice plane, while two parallel lines of broken bonds are being formed. The corresponding continuous model involves a pair of semi-infinite parallel cracks propagating along the $x-$axis.

Physically, the lattice model describes, for example, the process of pulling out a thread from fabric, where the propagation of the fault is accompanied by rupture of some of retaining bonds. Comparing the continuum and the lattice models we note a paradox occurring in the two-crack configuration in the infinite continuous plane. The finite energy release rate can be maintained at a constant level as the strip width and the force magnitude both tend to zero. Consequently, we arrive to the conclusion that a "fault" generating a finite energy release rate propagates under the action of a zero force. This paradox is resolved using the solution of the corresponding lattice problem.

It is also appropriate to mention another paradox associated with the continuum formulation in a different physical situation for lattice structures with edge or screw dislocations. This is linked to the classical Peierls-Nabarro model (Peierls (1940) and Nabarro (1947)), which includes, as a part of the mathematical formualtion, a hypersingular integral equation describing a displacement jump across the glide plane – within a continuum approach any non-zero constant stress, represented by a constant term in the right-hand side of the equation, is sufficient to move the dislocation all the way to infinity, which indeed contradicts the physical reality. On the other hand, the lattice provides the trapping effect and hence a positive value of the critical Peierls stress.

Lattice formulation for a double-crack fault

We consider a steady-state dynamic problem for a square uniform lattice where a chain of connected particles is extracted from the lattice by a remote force. The problem is normalized in such a way that the lattice spacing, stiffness of bonds and the mass of particles are all equal to unity. This can be thought of as a propagation of two parallel cracks. As before, the problem is assumed to be symmetric with respect to the $x-$axis, so that it is sufficient to consider the upper half-plane. For the first two layers, the

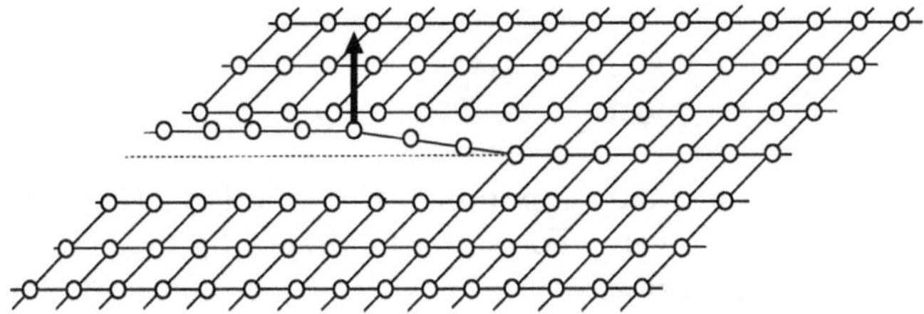

Figure 8. The lattice with the extracted chain. In the problem formulation, the force shown by the arrow is applied far from the crack front, at 'minus infinity'.

equations of motion are

$$\begin{aligned}\ddot{u}_{0,m} &= u_{0,m+1} + u_{0,m-1} + 2u_{1,m} - 4u_{0,m} - 2(u_{1,m} - u_{0,m})H(-\eta),\\ \ddot{u}_{1,m} &= u_{1,m+1} + u_{1,m-1} + u_{2,m} + u_{0,m} - 4u_{1,m} + (u_{1,m} - u_{0,m})H(-\eta).\end{aligned} \quad (70)$$

The first subscript stands for the number of the horizontal layer, whereas the second subscript denotes the number of the node within the layer. The Fourier transform with respect to $\eta = m - vt$ leads to the equations

$$\begin{aligned}(2+h^2)u_0^F - 2u_1^F &= -2(u_{1-} - u_{0-}),\\ u_0^F + (2+h^2-\lambda)u_1^F &= (u_{1-} - u_{0-}),\end{aligned} \quad (71)$$

where $\lambda = u_2^F/u_1^F$.

We use the notation

$$Q^F = (u_1 - u_0)^F. \quad (72)$$

In Mishuris *et al.* (2007), the problem is reduced to a functional equation of the Wiener-Hopf type. The homogeneous equation has the form

$$Q_+ + L(k)Q_- = 0 \quad (73)$$

with

$$L(k) = \frac{h^2(s+h)}{s(s^2+sh-2)}, \quad (74)$$

where

$$h = \sqrt{2(1-\cos k)^2 + (0 + ikv)^2}, \quad s = \sqrt{h^2+4}.$$

Note that

$$s + h \neq 0\,, \quad s^2 + sh - 2 \neq 0 \tag{75}$$

and hence real zeros and singular points of the function $L(k)$ coinsides with those of h and s, respectively. For subcritical speeds, $0 \leq v < 1$, and the kernel $L(k)$ has the properties

$$L(k) \sim \frac{1-v^2}{2}k^2 \quad (k \to 0)\,, \quad L(k) \to 1 \quad (k \to \pm\infty)\,, \quad \operatorname{Ind} L(k) = 0\,. \tag{76}$$

This function allows for the factorization

$$\begin{aligned} L(k) &= L(k)_+ L_-(k)\,, \\ L_\pm(k) &= \exp\left[\pm\frac{1}{2\pi\mathrm{i}}\int_{-\infty}^{\infty}\frac{\ln L(\xi)}{\xi - k}\,\mathrm{d}\xi\right] \quad \text{with} \quad \pm\Im k > 0\,. \end{aligned} \tag{77}$$

We shall also use the notation

$$R(v) = \exp\left[\frac{1}{\pi}\int_0^{\infty}\frac{\operatorname{Arg} L(k)}{k}\,\mathrm{d}k\right]\,. \tag{78}$$

When a remote force is applied at infinity to the central chain, the equation becomes nonhomogeneous. The nonhomogeneous equations is defined according to Slepyan (2002), Section 11.5.1, with the delta-function type term in the right-hand side of the factorized equation :

$$\frac{Q_+}{L_+} + L_- Q_- = \frac{C}{0 - \mathrm{i}k} + \frac{C}{0 + \mathrm{i}k}\,, \tag{79}$$

where C is a constant. Detailed analysis of the solution is given in Mishuris *et al.* (2007). It has been shown that

$$C = -\frac{P}{R(v)}\sqrt{\frac{1-v^2}{2}}\,, \tag{80}$$

where P is the remote force applied to the central chain at infinity. The local energy release rate G_0, related to the energy released via breakage of the bonds at the tip of the double-crack fault, and the global energy release rate G, associated with the homogenized problem, are defined by

$$G_0 = \frac{(1-v^2)P^2}{2R^2(v)}\,, \tag{81}$$

$$G = \frac{(1-v^2)P^2}{2}\,, \tag{82}$$

and the energy release ratio is

$$\frac{G_0}{G} = R^{-2}(v)\,. \tag{83}$$

As before, the difference, $G - G_0$, can be interpreted as the energy dissipation rate for the lattice with the double-crack fault.

The continuum model versus the lattice solution

The related continuum model concerns with the anti-plane shear of an infinite plane containing two parallel cracks M_+ and M_-, separated by the distance a and propagating with the speed v along the $x-$axis:

$$\begin{aligned} M_+ &= \{(x,y) : \eta = x - vt < 0, y = a\} \\ \text{and}\quad M_- &= \{(x,y) : \eta = x - vt < 0, y = 0\}. \end{aligned} \tag{84}$$

The displacement function $u(\eta, y)$ satisfies the equation

$$\alpha^2 \frac{\partial^2 u(\eta,y)}{\partial \eta^2} + \frac{\partial^2 u(\eta,y)}{\partial y^2} = 0, \tag{85}$$

where $\alpha^2 = 1 - v^2/c^2$, and c is the shear wave speed; to be consistent with the formulation of Section 0.4 we will use the normalization $c = 1$.

The crack faces are traction free, which corresponds to the homogeneous Neumann boundary conditions

$$\frac{\partial u(\eta,y)}{\partial y} = 0 \;\text{ when }\; (x,y) \in M_+ \;\text{ and }\; (x,y) \in M_-. \tag{86}$$

The conditions at infinity are defined to model a remote load, the force $P = Ca$, applied at 'minus infinity' to the semi-infinite strip between the cracks M_+ and M_-:

$$u(\eta,y) \sim -C\eta \;\text{ as }\; \eta \to -\infty \;\text{ and }\; 0 < y < a, \tag{87}$$

and

$$u(\eta,y) \sim -\frac{\alpha C a}{4\pi} \ln\left(\eta^2 + \alpha^2 y^2\right) + \text{const} \tag{88}$$

when $\eta^2 + y^2 \to \infty$ outside the strip between M_+ and M_-.

The energy release rate

The above problem allows for an explicit closed form solution, which uses a conformal map of $\mathbf{R}^2 \setminus (M_+ \cup M_-)$ into the upper half-plane (see Morse and Feshbach (1953), p. 229):

$$z - z_0 = A\Big(\frac{1}{2}w^2 - \ln w\Big),$$

where $A = -a/\pi, \quad z_0 = -A/2$.

The solution of the problem (85)–(88) is not differentiable at the vertices of the cracks M_+ and M_-, i.e. the components of ∇u are singular, $O(1/(\eta^2 + y^2)^{1/4})$ and $O(1/(\eta^2 + (y-a)^2)^{1/4})$, as expected for stress near the crack tips in continuous media.

In contrast with the lattice model, the continuous formulation yields that the energy is not radiated to infinity, and all the energy produced by the remote force is absorbed into the vertices of the bouble-crack fault. The total energy release rate evaluated for the double-crack fault can be represented in the form

$$\mathcal{E} = \mathcal{E}_f - \mathcal{E}_s - \mathcal{E}_k, \tag{89}$$

where $\mathcal{E}_f = -Pu' = P^2/a$ is the rate of the work of the remote external force applied to the strip between the cracks, $\mathcal{E}_s = \frac{1}{2}(u')^2 a = \frac{P^2}{2a}$ is the strain energy rate of the strip, and $\mathcal{E}_k = \frac{\dot{u}^2 a}{2} = \frac{P^2 v^2}{2a}$ is the rate of the kinetic energy of the strip. Then the energy release rate per unit length for each of the two cracks is

$$G_{crack} = \frac{1}{2}\mathcal{E} = \frac{P^2(1-v^2)}{4a}. \tag{90}$$

The limit case of $a \to 0$

Assuming that the crack propagates when the energy release rate reaches the critical value, i.e. $G_{crack} = G_c$, one can evaluate the crack speed in the form

$$v = \sqrt{1 - \frac{4aG_c}{P^2}}. \tag{91}$$

When $P > 2\sqrt{aG_c}$ the speed v increases monotonically with the increase of the external force P, and $v \to 1$ as $P \to \infty$.

The formulae for the solution include the distance a between the cracks, and it is formally possible to consider the limit when $a \to 0$. Physically, this would correspond to a double-crack fault discussed in Section 0.4, which was treated in the framework of the lattice model. Here we show that subject to a certain constraint on the applied load, the formal application of the continuum formulation would lead to a paradox.

First, we assume that $P = \text{const}$. Then, as $a \to 0$, the crack speed tends to the critical value, which is equal to 1.

If $P/\sqrt{a} \to 0$, then the energy release rate tends to zero as $a \to 0$, and the crack should stop.

It is also feasible to take $P/\sqrt{a} \to \mathcal{P} > 2\sqrt{G_c}$. Then the force P tends to zero, whereas the energy release rate tends to a non-zero constant sufficient to support the propagating cracks. This leads to a paradox inherent to the continuum model: in the limit when $a \to 0$, the double-crack fault can propagate, as a point singularity, under a nonzero energy flux but zero external force. The paradox is fully resolved by the lattice model, as shown in the earlier Section 0.4.

Bibliography

Brillouin, L., 1953. Wave Propagation in Periodic Structures , Dover, NY.

Cai, C.W., Liu, J.K., Yang, Y., 2005. Exact analysis of localized modes in two-dimensional bi-periodic mass-spring systems with a single disorder, *Journal of Sound and Vibration* 288, 307–320.

Fineberg, J, Gross, S.P., Marder, M., and Swinney, H.L., 1991. Instability in dynamic fracture. Phys. Rev. Lett. 67, 457-460.

Fineberg, J, Gross, S.P., Marder, M., and Swinney, H.L., 1992. Instability in the propagation of fast cracks. Phys Rev B 45, 5146-5154.

Heizler, S.I., Kessler, D.A., and Levine, H., 2002. Mode-I fracture in a nonlinear lattice with viscoelastic forces. Phys. Rev. E 66, 016126.

John, S., 1987. Strong localization of Photons in certain disordered dielectric superlattices, *Physical Review Letters* 58, 2486–2489.

Jensen, J.S., 2003. Phononic band gaps and vibrations in one- and two-dimensional mass-spring structures, *Journal of Sound and Vibration* 226, 1053–1078.

Kessler, D.A., 1999. Arrested cracks in nonlinear lattice models of brittle fracture. Phys. Rev. E 60, 7569-7571.

Kessler, D.A., 2000. Steady-state cracks in viscoelastic lattice models. II. Phys. Rev. E 61, 2348 - 2360.

Kessler, D.A., and Levine, H., 2001. Nonlinear lattice model of viscoelastic mode III fracture. Phys. Rev. E 63, 016118.

Kessler, D.A., Levine, H. 2003. Does the continuum theory of dynamic fracture work? Phys. Rev. E 68, 036118.

Kulakhmetova, S.A., Saraikin, V.A., and Slepyan, L., 1984. Plane problem of a crack in a lattice. Mech. of Solids 19, 101-108.

Maradudin, A.A., Montroll, E.W., and Weiss, G.H., 1963. *Theory of Lattice Dynamics in the Harmonic Approximation*, Academic Press.

Marder, M., and Liu, X, 1994. Instability in lattice fracture. Phys. Rev. E 50, 188-197.

Marder, M., Gross, S., 1995. Origin of crack tip instabilities. J. Mech. Phys. Solids 43, 1-48.

Marder, M., Fineberg, J., 1996. How things break. Physics Today 49, 1-12.

Martin, P.A., Discrete scattering theory: Greens function for a square lattice, 2006. *Wave Motion* 43, 619–629.

Martinsson, P.G., Movchan, A.B., 2003. Vibrations of lattice structures and phononic band gaps, *The Quarterly Journal of Mechanics and Applied Mathematics* 56, 45–64.

Mishuris, G.S., Movchan, A.B., Slepyan, L.I., 2007. Waves and fracture in an inhomogeneous lattice structure. *Waves in Random and Complex Media*, 17, 409-428.

Mishuris, G.S., Movchan, A.B., Slepyan, L.I., 2008. Dynamical extraction of a single chain from a discrete lattice. *J. Mech. Phys. Solids*, 56, 487-495.

Mishuris, G.S., Movchan, A.B., Slepyan, L.I., 2008a. Dynamics of a bridged crack in a discrete lattice. The Quarterly Journal of Mechanics and Applied Mathematics, 61, 151-160.

Movchan, A.B., Slepyan, L.I., 2007. Band gap Green's functions and localized oscillations. *Proceedings of The Royal Soc. London A*, 463, 2709-2727.

Nabarro F.R.N., 1947. Dislocations in a simple cubic lattice. *Proc. Phys. Soc. London,* 59: 256-272.

Peierls R., 1940. The size of a dislocation. *Proc. Phys. Soc. London,* 52: 34.

Poulton, C.G., McPhedran, R.C., Nicorovici, N.A., Botten, L.C., 2003. Localized Green's functions for a two-dimensional periodic material. In: Movchan, A.B. (Ed.), IUTAM Symposium on Asymptotics, Singularities and Homogenisation in Problems of Mechanics, Kluwer, Dordrecht, 181-190.

Sharon, E., Gross, S.P., and Fineberg, J., 1996. Energy dissipation in dynamic fracture. Phys. Rev. Lett. 76, 2117-2120.

Slepyan, L.I., 1981. Dynamics of a crack in a lattice. Sov. Phys. Dokl., 26, 538-540.

Slepyan, L.I., 2000. Dynamic factor in impact, phase transition and fracture. J. Mech. Phys. Solids 48, 927-960.

Slepyan, L.I., 2001a. Feeding and dissipative waves in fracture and phase transition. I. Some 1D structures and a square-cell lattice. J. Mech. Phys. Solids, 49(3), 25-67.

Slepyan, L.I., 2001b. Feeding and dissipative waves in fracture and phase transition. III. Triangular-cell lattice. J. Mech. Phys. Solids 49(12), 2839-2875.

Slepyan, L.I., 2002. Models and Phenomena in Fracture Mechanics. Springer, Berlin.

Slepyan, L.I., 2005. Crack in a material-bond lattice. J. Mech. Phys. Solids 53, 1295-1313.

Slepyan, L.I., and Ayzenberg-Stepanenko, M.V., 2002. Some surprising phenomena in weak-bond fracture of a triangular lattice. J. Mech. Phys. Solids 50(8), 1591-1625.

Slepyan, L.I., and Ayzenberg-Stepanenko, M.V., 2004. Localized transition waves in bistable-bond lattices. J. Mech. Phys. Solids 52, 1447-1479.

Slepyan, L.I., and Ayzenberg-Stepanenko, M.V., 2006. Crack dynamics in nonlinear lattices. Int. J. Fracture 140 (1-4), 235-242.

Slepyan, L.I., Ayzenberg-Stepanenko, M., 2008. Resonant-Frequency Primitive Waveforms and Star Waves in Lattices, *Journal of Sound and Vibration*, 313, 812-821.

Yablonovitch, E., 1987. Inhibited spontaneous emission in solid-state physics and electronics, *Physical Review Letters* 58, 2059–2062.

Yablonovitch, E., 1993. Photonic band-gap crystals, *Journal of Physics: Condensed Matter* 5, 2443–2460.

A 3-D brick Cosserat Point Element (CPE) for nonlinear elasticity

M. Jabareen[*,†] and M. B. Rubin[**]
Institute of Mechanical Systems
Department of Mechanical Engineering ETH Zentrum, 8092 Zurich, Switzerland
Email: mahmood.jabareen@imes.mavt.ethz.ch
[†] Faculty of Civil and Environmental Engineering
Technion - Israel Institute of Technology, 32000 Haifa, Israel
Faculty of Mechanical Engineering
Technion - Israel Institute of Technology, 32000 Haifa, Israel
Email: mbrubin@tx.technion.ac.il

1 Introduction

A nonlinear hyperelastic elastic material has one of the simplest constitutive equations because the stress response is determined algebraically by derivatives of a strain energy function. However, the nonlinear partial differential equations which describe the deformation of an elastic material are intractable analytically for most problems. Therefore, numerical methods are essential to obtain solutions of realistic problems.

Examination of the commercial programs ABAQUS, ADINA, ANSYS and the academic code FEAP reveals that the user has to choose from a list of different hyperlastic elements (see Table 1.1). This list includes element formulations based on full integration (Q1), full integration of distortion and reduced integration of volume (Q1P0), reduced integration with various types of hourglass controls, hybrid methods, incompatible modes and enhanced strains. The reason for this extensive list is that no single element performs well for all element geometries, levels of compressibility and under all loading conditions. In particular, it is well known that, within the context of the Bubnov-Galerkin approach based on a tri-linear approximation of the displacement field, full integration of the constitutive equations leads to an element response which exhibits locking for bending dominated response of thin structures (shells and rods) with poor element aspect ratios and for nearly incompressible materials. Two main modifications of the element formulations have been proposed to overcome these problems. One modification uses reduced integration with hourglass control (e.g. Belytschko et al. (1984); Hutter et al. (2000); Reese and Wriggers (1996); Reese and Wriggers (2000); Reese et al. (2000)) and the second modification uses enhanced strains or incompatible modes (e.g. Simo and Rifai (1990); Simo and Armero (1992), and Simo et al. (1993)).Moreover, the enhanced strain and incompatible mode elements exhibit hourglass instabilities in regions of high compression combined with bending (e.g. Reese and Wriggers (1996); Reese and Wriggers (2000); Jabareen and Rubin (2007a), and Jabareen and Rubin (2007b)).

Often the person who wants to solve a specific problem using a hyperelastic constitutive equations typically does not know which of the element formulations in the code is best suited for the specific problem. Sometimes a problem can be sufficiently complicated that none of these elements can provide accurate predictions for the deformation fields in all regions of the problem. In this sense the existing element formulations are not user friendly.

Therefore, there appears to be a need for a robust user friendly element formulation that can be reliably used for all applications. The 3-D brick Cosserat Point Element (CPE) is a new element technology that is based on the theory of a Cosserat point (Rubin (1985a); Rubin (1985b); Rubin (1995), and Rubin (2000)) and which has been proven (Nadler and Rubin (2003); Jabareen and Rubin (2007a); Jabareen and Rubin (2007b); Jabareen and Rubin (2007c), and Jabareen and Rubin (2008a)) to be such a robust user friendly element for nonlinear elasticity. In particular, it does not exhibit unphysical locking or hourglassing for thin structures or nearly incompressible material response.

2 Basic tensor operations

Before developing the equations for a CPE it is useful to review some tensor operations. Basic knowledge of index notation and simple vector operations is assumed and more details of tensor operations can be found in (Rubin (2000)).

Tensor product

Let $\{\mathbf{a}_i,\ \mathbf{b}_i\}$ $(i = 1, 2, 3, 4)$ be sets of vectors in three-dimensional space. Then, the tensor product is denoted by the symbol $\otimes$ and the tensor product $\mathbf{a}_1 \otimes \mathbf{a}_2$ of two vectors is defined by its operation on another vector $\mathbf{b}_1$, such that

$$\begin{aligned}(\mathbf{a}_1 \otimes \mathbf{a}_2)\mathbf{b}_1 &= \mathbf{a}_1(\mathbf{a}_2 \bullet \mathbf{b}_1) = (\mathbf{a}_2 \bullet \mathbf{b}_1)\mathbf{a}_1 \quad , \\ \mathbf{b}_1(\mathbf{a}_1 \otimes \mathbf{a}_2) &= (\mathbf{b}_1 \bullet \mathbf{a}_1)\mathbf{a}_2 \end{aligned} \tag{2.1}$$

Table 1.1. List of elements tested in the programs ABAQUS, ADINA, ANSYS and FEAP.

Short Name	Program	Element Name	Options
ABAQUS-1	ABAQUS	C3D8^{S}	full integration ($\bar{\mathrm{B}}$ method)
ABAQUS-2	ABAQUS	C3D8R^{S}	reduced integration hourglass control: enhanced
ABAQUS-3	ABAQUS	C3D8R^{S}	reduced integration hourglass control: sti ness
ABAQUS-4	ABAQUS	C3D8RH^{S}	hybrid formulation reduced integration hourglass control: enhanced
ABAQUS-5	ABAQUS	C3D8RH^{S}	hybrid formulation reduced integration hourglass control: sti ness
ABAQUS-6	ABAQUS	C3D8I^{S}	incompatible modes
ABAQUS-7	ABAQUS	C3D8IH^{S}	hybrid formulation incompatible modes
ADINA-1	ADINA	3D Solid (8 Nodes)	full integration
ADINA-2	ADINA	3D Solid (8 Nodes)	incompatible modes
ANSYS-1	ANSYS	Solid185	pure displacement with full integration
ANSYS-2	ANSYS	Solid185	pure displacement with reduced integration and hourglass control
ANSYS-3	ANSYS	Solid185	pure displacement with enhanced strains
ANSYS-4	ANSYS	Solid185	pure displacement with simplified enhanced strains
ANSYS-5	ANSYS	Hyper86	full integration
ANSYS-6	ANSYS	Hyper86	full integration for shear and reduced integration for volume
FEAP-1	FEAP	Solid Fini-Disp-8	full integration
FEAP-2	FEAP	Solid Fini-Mixe-8	mixed formulation
FEAP-3	FEAP	Solid Fini-Enha-8	enhanced strains

where $(\mathbf{a}_2 \bullet \mathbf{b}_1)$ denotes the scalar product between the two vectors $\mathbf{a}_2$, $\mathbf{b}_1$. A scalar is called a zero order tensor and a vector is called a first order tensor. The quantity $(\mathbf{a}_1 \otimes \mathbf{a}_2)$ is called a second order tensor because it is linear operator which maps the space of vectors onto a tensor of one order lower than itself (i.e. a first order tensor). That is to say that the result of $(\mathbf{a}_1 \otimes \mathbf{a}_2)$ operating on a vector is a vector. In general, the result when $\mathbf{b}_1$ is placed to the right of $(\mathbf{a}_1 \otimes \mathbf{a}_2)$ is different than when it is placed to the left of $(\mathbf{a}_1 \otimes \mathbf{a}_2)$.

The tensor product can be used to develop higher order tensors by creating a string of vectors, separated by tensor products. For example, the quantity $(\mathbf{a}_1 \otimes \mathbf{a}_2 \otimes \mathbf{a}_3 \otimes \mathbf{a}_4)$ is a fourth order tensor which satisfies the conditions

$$\begin{aligned}(\mathbf{a}_1 \otimes \mathbf{a}_2 \otimes \mathbf{a}_3 \otimes \mathbf{a}_4)\mathbf{b}_1 &= (\mathbf{a}_4 \bullet \mathbf{b}_1)(\mathbf{a}_1 \otimes \mathbf{a}_2 \otimes \mathbf{a}_3) \ , \\ \mathbf{b}_1(\mathbf{a}_1 \otimes \mathbf{a}_2 \otimes \mathbf{a}_3 \otimes \mathbf{a}_4) &= (\mathbf{b}_1 \bullet \mathbf{a}_1)(\mathbf{a}_2 \otimes \mathbf{a}_3 \otimes \mathbf{a}_4)\end{aligned} \tag{2.2}$$

Juxtaposition of two tensors

The operation of juxtaposition is used when two tensor are placed next to each other. For example

$$\begin{aligned}(\mathbf{a}_1 \otimes \mathbf{a}_2)(\mathbf{b}_1 \otimes \mathbf{b}_2) &= \mathbf{a}_1 \otimes (\mathbf{a}_2 \bullet \mathbf{b}_1)\mathbf{b}_2 = (\mathbf{a}_2 \bullet \mathbf{b}_1)(\mathbf{a}_1 \otimes \mathbf{b}_2) \ , \\ (\mathbf{b}_1 \otimes \mathbf{b}_2)(\mathbf{a}_1 \otimes \mathbf{a}_2) &= (\mathbf{b}_2 \bullet \mathbf{a}_1)(\mathbf{b}_1 \otimes \mathbf{a}_2) \neq (\mathbf{a}_1 \otimes \mathbf{a}_2)(\mathbf{b}_1 \otimes \mathbf{b}_2)\end{aligned} \tag{2.3}$$

In particular, note that the operation of juxtaposition is not commutative so that the two tensors in (2.3a, 2.3b) are not necessarily equal. Moreover, it is noted that the operation of juxtaposition involves the scalar product of only one vector from each of the tensors. It will be shown later that this operation yields results that are the same as standard multiplication of matrices.

Dot product of two tensors

The scalar product or dot product of two vectors is a positive definite operator which is defined so that the dot product of a vector with itself is positive as long as the vector is nonzero. The dot product of two tensors is also defined as a positive definite operator. Specifically, it is defined so that

$$\begin{aligned}&(\mathbf{a}_1 \otimes \mathbf{a}_2) \bullet (\mathbf{b}_1 \otimes \mathbf{b}_2) = (\mathbf{a}_1 \bullet \mathbf{b}_1)(\mathbf{a}_2 \bullet \mathbf{b}_2) = (\mathbf{b}_1 \otimes \mathbf{b}_2) \bullet (\mathbf{a}_1 \otimes \mathbf{a}_2) \ , \\ &(\mathbf{a}_1 \otimes \mathbf{a}_2 \otimes \mathbf{a}_3 \otimes \mathbf{a}_4) \bullet (\mathbf{b}_1 \otimes \mathbf{b}_2) = (\mathbf{a}_3 \bullet \mathbf{b}_1)(\mathbf{a}_4 \bullet \mathbf{b}_2)(\mathbf{a}_1 \otimes \mathbf{a}_2) \ , \\ &(\mathbf{b}_1 \otimes \mathbf{b}_2) \bullet (\mathbf{a}_1 \otimes \mathbf{a}_2 \otimes \mathbf{a}_3 \otimes \mathbf{a}_4) = (\mathbf{b}_1 \bullet \mathbf{a}_1)(\mathbf{b}_2 \bullet \mathbf{a}_2)(\mathbf{a}_3 \otimes \mathbf{a}_4) \\ &\neq (\mathbf{a}_1 \otimes \mathbf{a}_2 \otimes \mathbf{a}_3 \otimes \mathbf{a}_4) \bullet (\mathbf{b}_1 \otimes \mathbf{b}_2) \ , \\ &(\mathbf{a}_1 \otimes \mathbf{a}_2 \otimes \mathbf{a}_3 \otimes \mathbf{a}_4) \bullet (\mathbf{b}_1 \otimes \mathbf{b}_2 \otimes \mathbf{b}_3 \otimes \mathbf{b}_4) \\ &= (\mathbf{a}_1 \bullet \mathbf{b}_1)(\mathbf{a}_2 \bullet \mathbf{b}_2)(\mathbf{a}_3 \bullet \mathbf{b}_3)(\mathbf{a}_4 \bullet \mathbf{b}_4)\end{aligned} \tag{2.4}$$

In particular, note that the dot product of two tensors has the order of the difference of the orders of the tensors (e.g. the dot product of two tensors of the same order is a scalar and the dot product of a tensor of fourth order with a tensor of second order is a tensor of second order).

Transpose of a tensor

The transpose of a tensor is obtained by interchanging the order of the vectors associated with the order of the transpose operator. For example, the right transpose (denoted by a superposed T) and the left transpose (denoted by a superposed LT) are defined so that

$$\begin{aligned}(\mathbf{a}_1 \otimes \mathbf{a}_2)^T &= (\mathbf{a}_2 \otimes \mathbf{a}_1) \ , \\ {}^{LT}(\mathbf{a}_1 \otimes \mathbf{a}_2) &= (\mathbf{a}_2 \otimes \mathbf{a}_1) \ , \\ (\mathbf{a}_1 \otimes \mathbf{a}_2 \otimes \mathbf{a}_3 \otimes \mathbf{a}_4)^T &= (\mathbf{a}_1 \otimes \mathbf{a}_2) \otimes (\mathbf{a}_4 \otimes \mathbf{a}_3) \ , \\ {}^{LT}(\mathbf{a}_1 \otimes \mathbf{a}_2 \otimes \mathbf{a}_3 \otimes \mathbf{a}_4) &= (\mathbf{a}_2 \otimes \mathbf{a}_1) \otimes (\mathbf{a}_3 \otimes \mathbf{a}_4)\end{aligned} \tag{2.5}$$

Notice that the operators T and LT change the order of the two vectors closest to the operator. It is also possible to define higher order transpose operations like $T(2)$ and $LT(2)$ which apply to pairs of two vectors, such that

$$\begin{aligned}(\mathbf{a}_1 \otimes \mathbf{a}_2 \otimes \mathbf{a}_3 \otimes \mathbf{a}_4)^{T\ 2)} &= (\mathbf{a}_3 \otimes \mathbf{a}_4) \otimes (\mathbf{a}_1 \otimes \mathbf{a}_2) \ , \\ &= {}^{LT\ 2)}(\mathbf{a}_1 \otimes \mathbf{a}_2 \otimes \mathbf{a}_3 \otimes \mathbf{a}_4)\end{aligned} \tag{2.6}$$

General tensors

Although the tensor $(\mathbf{a}_1 \otimes \mathbf{a}_2)$ is a second order tensor it is not a general second order tensor. In order to discuss general tensors it is convenient to first consider tensors referred to a rectangular Cartesian triad $\mathbf{e}_i \, (i = 1, 2, 3)$ of constant orthonormal vectors

$$\mathbf{e}_i \bullet \mathbf{e}_j = \delta_{ij} \tag{2.7}$$

where δ_{ij} denotes the Kroneker delta

$$\delta_{ij} = 1 \quad for \quad i = j \quad and \quad \delta_{ij} = 0 \quad for \quad i \neq j \tag{2.8}$$

It is well known that the vectors $\mathbf{e}_i$ form a complete set of base vectors that span the space of three-dimensional vectors so that an arbitrary vector $\mathbf{v}$ can be expressed in terms of its components v_i relative to $\mathbf{e}_i$, such that

$$\begin{aligned}v_i &= \mathbf{v} \bullet \mathbf{e}_i \ , \\ \mathbf{v} &= v_i \mathbf{e}_i\end{aligned} \tag{2.9}$$

where the usual summation convention is used over repeated indices, which take the values ($i = 1, 2, 3$). In a similar manner it is possible to define a set of nine orthonormal base tensors ($\mathbf{e}_i \otimes \mathbf{e}_j$) that span the space of second order tensors with

$$(\mathbf{e}_i \otimes \mathbf{e}_j) \bullet (\mathbf{e}_m \otimes \mathbf{e}_n) = \delta_{im}\delta_{jn} \tag{2.10}$$

Then, an arbitrary second order tensor $\mathbf{T}$ can be expressed in terms of its components T_{ij} relative to $\mathbf{e}_i$, such that

$$\begin{aligned} T_{ij} &= \mathbf{T} \bullet (\mathbf{e}_i \otimes \mathbf{e}_j) \ , \\ \mathbf{T} &= T_{ij}(\mathbf{e}_i \otimes \mathbf{e}_j) \end{aligned} \tag{2.11}$$

Also, an arbitrary fourth order tensor $\mathbf{T}$ can be expressed in terms of its components T_{ijmn} relative to $\mathbf{e}_i$, such that

$$\begin{aligned} T_{ijmn} &= \mathbf{T} \bullet (\mathbf{e}_i \otimes \mathbf{e}_j \otimes \mathbf{e}_m \otimes \mathbf{e}_n) \ , \\ \mathbf{T} &= T_{ijmn}(\mathbf{e}_i \otimes \mathbf{e}_j \otimes \mathbf{e}_m \otimes \mathbf{e}_n) \end{aligned} \tag{2.12}$$

In particular, it is noted that a general second order tensor has $3^2 = 9$ independent components and a general fourth order tensor has $3^4 = 81$ independent components.

Representation of tensors with respect to curvilinear coordinates
Within the context of the CPE theory it is convenient to express some tensors using the same symbol as is typically used in the three-dimensional theory. Thus, in order to distinguish between these quantities a superposed (*) is used for the three-dimensional quantity. For example, the position vector of a material point in the deformed present configuration associated with the three-dimensional theory is denoted by $\mathbf{x}^*$ instead of $\mathbf{x}$. The rectangular Cartesian base vectors $\mathbf{e}_i$ are special in that they are constants which are independent of the coordinates x_i^*. Consequently, the position vector $\mathbf{x}^*$ can be expressed in the form

$$\mathbf{x}^* = x_i^* \mathbf{e}_i \tag{2.13}$$

so that the base vectors $\mathbf{e}_i$ can be determined by the equations

$$\mathbf{e}_i = \frac{\partial \mathbf{x}^*}{\partial x_i^*} \tag{2.14}$$

For general curvilinear coordinates the position vector $\mathbf{x}^*$ is a function of three convected coordinates θ^i (i=1,2,3) and time t, such that

$$\mathbf{x}^* = \mathbf{x}^*(\theta^i, t) \tag{2.15}$$

The convected coordinates θ^i have constant values for a specified material point and the need for distinguishing between subscripts and superscripts will become apparent. The covariant base vectors $\mathbf{g}_i$ defined by

$$\mathbf{g}_i = \mathbf{x}^*_{,i} = \frac{\partial \mathbf{x}^*}{\partial \theta^i} \tag{2.16}$$

are generalizations of the base vectors $\mathbf{e}_i$ in (2.14). Here, and throughout the text, a comma is used to denote partial differentiation with respect to the coordinates θ^i. Also, it is noted that the mapping (2.15) is limited so that it is one-to-one with $\mathbf{g}_i$ being linearly independent vectors

$$g^{1/2} = \mathbf{g}_1 \times \mathbf{g}_2 \bullet \mathbf{g}_3 > 0 \tag{2.17}$$

that span the three-dimensional space. The main difference between $\mathbf{g}_i$ and $\mathbf{e}_i$ is that $\mathbf{g}_i$ can depend on the coordinates θ^i. This has an important influence on expressions related to the gradient and divergence operators. Moreover, since θ^i do not necessarily have the units of length (i.e. the angle in cylindrical polar coordinates), $\mathbf{g}_i$ need not be unitless.

Since $\mathbf{g}_i$ are linearly independent it is possible to define reciprocal vectors $\mathbf{g}^i$ (also called contravariant base vectors) by the expressions

$$\begin{aligned} \mathbf{g}^1 &= g^{-1/2}\mathbf{g}_2 \times \mathbf{g}_3 \ , \quad \mathbf{g}^2 = g^{-1/2}\mathbf{g}_3 \times \mathbf{g}_1 \ , \\ \mathbf{g}^3 &= g^{-1/2}\mathbf{g}_1 \times \mathbf{g}_2 \end{aligned} \tag{2.18}$$

such that

$$\mathbf{g}^i \bullet \mathbf{g}_j = \delta^i_j \tag{2.19}$$

where δ^i_j is the Kronecker delta symbol. Now, an arbitrary vector $\mathbf{v}$ can be expressed in terms of its covariant components v_i or its contravariant components v^i

$$\begin{aligned} v_i &= \mathbf{v} \bullet \mathbf{g}_i \ , \quad v^i = \mathbf{v} \bullet \mathbf{g}^i \ , \\ \mathbf{v} &= v_i \mathbf{g}^i = v^i \mathbf{g}_i \end{aligned} \tag{2.20}$$

Similarly, an arbitrary second order tensor $\mathbf{T}$ can be expressed in terms of its covariant components T_{ij}, its contravariant components T^{ij}, or its mixed components $T^i_{\ j}, T_i^{\ j}$

$$\begin{aligned} T_{ij} &= \mathbf{T} \bullet (\mathbf{g}_i \otimes \mathbf{g}_j) \ , \quad T^{ij} = \mathbf{T} \bullet (\mathbf{g}^i \otimes \mathbf{g}^j) \ , \\ T^i_{\ j} &= \mathbf{T} \bullet (\mathbf{g}^i \otimes \mathbf{g}_j) \ , \quad T_i^{\ j} = \mathbf{T} \bullet (\mathbf{g}_i \otimes \mathbf{g}^j) \ , \\ \mathbf{T} &= T_{ij}(\mathbf{g}^i \otimes \mathbf{g}^j) = T^{ij}(\mathbf{g}_i \otimes \mathbf{g}_j) \ , \\ &= T^i_{\ j}(\mathbf{g}_i \otimes \mathbf{g}^j) = T_i^{\ j}(\mathbf{g}^i \otimes \mathbf{g}_j) \end{aligned} \tag{2.21}$$

In particular, notice that the summation connects covariant components with contravariant base vectors or contravariant components with covariant base vectors. Furthermore, it can be shown that the unit second order tensor $\mathbf{I}$ can be expressed in the forms

$$\mathbf{I} = (\mathbf{g}_i \otimes \mathbf{g}^i) = (\mathbf{g}^i \otimes \mathbf{g}_i) \tag{2.22}$$

Referential description

In continuum mechanics it is sometimes convenient to introduce a stress-free reference configuration. Specifically, the material point in the reference configuration that is associated with the position $\mathbf{x}^*$ in the present configuration is denoted by $\mathbf{X}^*$

$$\mathbf{X}^* = \mathbf{X}^*(\theta^i) \tag{2.23}$$

and is independent of time t. It then follows that the associated covariant base vectors $\mathbf{G}_i$, and contravariant base vectors $\mathbf{G}^i$ are defined by expressions similar to (2.16)-(2.19)

$$\begin{aligned}
&\mathbf{G}_i = \mathbf{X}^*_{,i} \ , \quad G^{1/2} = \mathbf{G}_1 \times \mathbf{G}_2 \bullet \mathbf{G}_3 > 0 \ , \quad \mathbf{G}^i \bullet \mathbf{G}_j = \delta^i_j \ , \\
&\mathbf{G}^1 = G^{-1/2} \mathbf{G}_2 \times \mathbf{G}_3 \ , \quad \mathbf{G}^2 = G^{-1/2} \mathbf{G}_3 \times \mathbf{G}_1 \ , \\
&\mathbf{G}^3 = G^{-1/2} \mathbf{G}_1 \times \mathbf{G}_2
\end{aligned} \tag{2.24}$$

Also, the unit tensor $\mathbf{I}$ can be written in the alternative forms

$$\mathbf{I} = (\mathbf{G}_i \otimes \mathbf{G}^i) = (\mathbf{G}^i \otimes \mathbf{G}_i) \tag{2.25}$$

Gradient operator

The gradient of a tensor $\mathbf{T}$ relative to the reference position $\mathbf{X}^*$ is denoted by $Grad^*\mathbf{T}$ and the gradient of $\mathbf{T}$ relative to the present position $\mathbf{x}^*$ is denoted by $grad^*\mathbf{T}$, which are defined by

$$Grad^*\mathbf{T} = \frac{\partial \mathbf{T}}{\partial \mathbf{X}^*} = \mathbf{T}_{,i} \otimes \mathbf{G}^i \ , \quad grad^*\mathbf{T} = \frac{\partial \mathbf{T}}{\partial \mathbf{x}^*} = \mathbf{T}_{,i} \otimes \mathbf{g}^i \tag{2.26}$$

Divergence operator

The divergence of a tensor $\mathbf{T}$ relative to the reference position $\mathbf{X}^*$ is denoted by $Div^*\mathbf{T}$ and the divergence of $\mathbf{T}$ relative to the present position $\mathbf{x}^*$ is denoted by $div^*\mathbf{T}$, which are defined by

$$Div^*\mathbf{T} = \mathbf{T}_{,i} \bullet \mathbf{G}^i \ , \quad div^*\mathbf{T} = \mathbf{T}_{,i} \bullet \mathbf{g}^i \tag{2.27}$$

The divergence operators can be simplified by differentiating (2.17), (2.18) and (2.24) to prove the identities

$$(G^{1/2}\mathbf{G}^i)_{,i} = \mathbf{0} \ , \quad (g^{1/2}\mathbf{g}^i)_{,i} = \mathbf{0} \tag{2.28}$$

Then, $Div^*\mathbf{T}$ and $div^*\mathbf{T}$ can be expressed in the alternative forms

$$Div^*\mathbf{T} = G^{\;1/2}(G^{1/2}\mathbf{T}\mathbf{G}^i)_{\,i} \quad , \quad div^*\mathbf{T} = g^{\;1/2}(g^{1/2}\mathbf{T}\mathbf{g}^i)_{\,i} \tag{2.29}$$

These forms are simpler than (2.27) because the derivative of $\mathbf{T}$ includes derivatives of the components of $\mathbf{T}$ as well as derivatives of each of the base vectors. For example, if $\mathbf{T}$ is a second order tensor then

$$\begin{aligned}\mathbf{T}_{\,i} &= [T_{mn}(\mathbf{g}^m \otimes \mathbf{g}^n)]_{\,i} \\ &= T_{mn\,i}(\mathbf{g}^m \otimes \mathbf{g}^n) + T_{mn}(\mathbf{g}^m_{\,i} \otimes \mathbf{g}^n) + T_{mn}(\mathbf{g}^m \otimes \mathbf{g}^n_{\,i})\end{aligned} \tag{2.30}$$

whereas, the expressions in (2.29) are based on derivatives of vectors.

3 Some kinematic measures in continuum mechanics

In continuum mechanics the material point $\mathbf{X}^*$ in the fixed reference configuration mapped to the location $\mathbf{x}^*$ in the deformed present configuration by the expression (2.15). The absolute velocity $\mathbf{v}^*$ of the material point is obtained by

$$\mathbf{v}^* = \dot{\mathbf{x}}^*(\theta^i, t) = \frac{\partial \mathbf{x}^*(\theta^i, t)}{\partial t} \tag{3.1}$$

where the material time derivative is denoted by a superposed dot $(\cdot)$ which indicates partial differentiation with respect to time t holding the convected coordinates θ^i constants.

The deformation gradient $\mathbf{F}^*$ is a two-point tensor that maps material line element $\mathbf{dX}^*$ in the reference configuration to material line elements $\mathbf{dx}^*$ in the present configuration

$$\mathbf{dx}^* = \mathbf{F}^*\mathbf{dX}^* \quad , \quad \mathbf{F}^* = \frac{\partial \mathbf{x}^*}{\partial \mathbf{X}^*} = Grad^*\mathbf{x}^* \tag{3.2}$$

Using the results in Section 2 it can be shown that $\mathbf{F}^*$ can be expressed in terms of the base vectors by

$$\mathbf{F}^* = \mathbf{g}_i \otimes \mathbf{G}^i \tag{3.3}$$

Moreover, the dilatation J^*

$$J^* = det(\mathbf{F}^*) = \frac{g^{1/2}}{G^{1/2}} = \frac{dv^*}{dV^*} \tag{3.4}$$

is a pure measure of volume change since the element of volume dV^* in the reference configuration and the element of volume dv^* in the present configuration are given by

$$dV^* = G^{1/2}d\theta^1 d\theta^2 d\theta^3 \quad , \quad dv^* = g^{1/2}d\theta^1 d\theta^2 d\theta^3 \tag{3.5}$$

Next, taking the material derivative of $\mathbf{F}^*$ and using the fact that $\mathbf{G}^i$ are independent of time it can be shown that

$$\dot{\mathbf{F}}^* = \dot{\mathbf{g}}_i \otimes \mathbf{G}^i = \mathbf{v}^*_{,i} \otimes \mathbf{G}^i = \mathbf{L}^* \mathbf{g}_i \otimes \mathbf{G}^i = \mathbf{L}^* \mathbf{F}^* \tag{3.6}$$

where $\mathbf{L}^*$ is the velocity gradient

$$\mathbf{L}^* = \frac{\partial \mathbf{v}^*}{\partial \mathbf{x}^*} = \mathbf{v}^*_{,j} \otimes \mathbf{g}^j \tag{3.7}$$

The velocity gradient $\mathbf{L}^*$ separates into its symmetric part $\mathbf{D}^*$, called the rate of deformation tensor, and its skew-symmetric part $\mathbf{W}^*$, called the spin tensor, such that

$$\begin{aligned} \mathbf{L}^* &= \mathbf{D}^* + \mathbf{W}^* \ , \\ \mathbf{D}^* &= \frac{1}{2}(\mathbf{L}^* + \mathbf{L}^{*T}) = \mathbf{D}^{*T} \ , \\ \mathbf{W}^* &= \frac{1}{2}(\mathbf{L}^* \quad \mathbf{L}^{*T}) = \quad \mathbf{W}^{*T} \end{aligned} \tag{3.8}$$

Moreover, it can be shown that the material derivative of the dilatation J^* is given by

$$\dot{J}^* = J^* \mathbf{D}^* \bullet \mathbf{I} \tag{3.9}$$

4 Balance laws in curvilinear coordinates

Let P denote the current material region of space occupied by a body and let ∂P be its smooth closed boundary. Also, let P_0 be the material region occupied by the same body in its fixed reference configuration with ∂P_0 being its smooth closed boundary. Then, the balance laws of the purely mechanical theory can be expressed as the conservation of mass

$$\int_P \rho^* dv^* = \int_P \rho_0^* dV^* \tag{4.1}$$

the balance of linear momentum

$$\frac{d}{dt} \int_P \rho^* \mathbf{v}^* dv^* = \int_P \rho^* \mathbf{b}^* dv^* + \int_{\partial P} \mathbf{t}^* da^* \tag{4.2}$$

and the balance of angular momentum about the fixed origin

$$\frac{d}{dt} \int_P \mathbf{x}^* \times \rho^* \mathbf{v}^* dv^* = \int_P \mathbf{x}^* \times \rho^* \mathbf{b}^* dv^* + \int_{\partial P} \mathbf{x}^* \times \mathbf{t}^* da^* \tag{4.3}$$

In these expressions, ρ^* is the current mass density, ρ_0^* is its reference value, $\mathbf{b}^*$ is the body force per unit mass, $\mathbf{t}^*$ is the surface traction and da^* is the element of area in the present configuration. Also, it is recalled that the traction vector is related to the Cauchy stress tensor $\mathbf{T}^*$ and the unit outward normal vector $\mathbf{n}^*$ to ∂P by the expression

$$\mathbf{t}^* = \mathbf{T}^*\mathbf{n}^* \tag{4.4}$$

Next, using (3.5) and the divergence theorem in the form

$$\int_P div^* \mathbf{A} dv^* = \int_{\partial P} \mathbf{A}\mathbf{n}^* da^* \tag{4.5}$$

the local forms of the conservation of mass and the balance of linear momentum become

$$m^* = \rho^* g^{1/2} = \rho_0^* G^{1/2} = m^*(\theta^i) \quad , \quad m^* \dot{\mathbf{v}}^* = m^* \mathbf{b}^* + \mathbf{t}^{*j}_{\ ,j} \tag{4.6}$$

where use has been made of (2.28) and the three vectors $\mathbf{t}^{*i}$ are defined by

$$\mathbf{t}^{*i} = g^{1/2} \mathbf{T}^* \mathbf{g}^i \tag{4.7}$$

Also, using the balance laws (4.6) the reduced form of the balance of angular momentum requires the Cauchy stress tensor to be symmetric

$$\mathbf{T}^{*T} = \mathbf{T}^* \tag{4.8}$$

Within the context of the purely mechanical theory it is convenient to define the rate of work W done on the body, the kinetic energy K and the strain energy U of the body by the expressions

$$\mathsf{W} = \int_P \rho^* \mathbf{b}^* \bullet \mathbf{v}^* dv^* + \int_{\partial P} \mathbf{t}^* \bullet \mathbf{v}^* da^* \quad ,$$

$$\mathsf{K} = \int_P \frac{1}{2} \rho^* \mathbf{v}^* \bullet \mathbf{v}^* dv^* \quad , \quad \mathsf{U} = \int_P \rho^* \Sigma^* dv^* \tag{4.9}$$

where Σ^* is the strain energy function per unit mass. Then, the rate of material dissipation $\ ^*$ per unit present volume can be defined by

$$\int_P {}^* dv^* = \mathsf{W} \quad \dot{\mathsf{K}} \quad \dot{\mathsf{U}} \geq 0 \tag{4.10}$$

and is required to be non-negative. Next, using the balance laws (4.6) and (4.8) it can be shown that

$${}^* = \mathbf{T}^* \bullet \mathbf{D}^* \quad \rho^* \dot{\Sigma}^* \geq 0 \tag{4.11}$$

For a nonlinear elastic solid the strain energy depends on the deformation gradient $\mathbf{F}^*$ through the deformation tensor $\mathbf{C}^*$

$$\Sigma^* = \Sigma^*(\mathbf{C}^*) \quad , \quad \mathbf{C}^* = \mathbf{F}^{*T}\mathbf{F}^* \tag{4.12}$$

the stress $\mathbf{T}^*$ is assumed to be independent of deformation rate and the rate of dissipation $\;^*$ vanishes. These assumptions lead to the result that the stress is given by the hyperelastic constitutive equation

$$\mathbf{T}^* = 2\rho^*\mathbf{F}^*\frac{\partial \Sigma^*}{\partial \mathbf{C}^*}\mathbf{F}^{*T} \tag{4.13}$$

Using the work of Flory (1961) it is possible to separate the effects of dilation from distortion. Specifically, the dilation J^* defined by

$$J^* = det(\mathbf{F}^*) \tag{4.14}$$

is a pure measure of volume change and the symmetric, unimodular tensor $\mathbf{B}^{*\prime}$ defined by

$$\mathbf{B}^{*\prime} = J^{* \quad 2/3}\mathbf{B}^* \quad , \quad \mathbf{B}^* = \mathbf{F}^*\mathbf{F}^{*T} \quad , \quad det(\mathbf{B}^{*\prime}) = 1 \tag{4.15}$$

is a pure measure of distortional deformation. It therefore, follows that $\mathbf{B}^{*\prime}$ has only two nontrivial invariants, which can be defined by

$$\;^*_1 = \mathbf{B}^{*\prime} \bullet \mathbf{I} \quad , \quad \;^*_2 = \mathbf{B}^{*\prime} \bullet \mathbf{B}^{*\prime} \tag{4.16}$$

Thus, for an elastically isotropic material the strain energy function Σ^* can be expressed in the form

$$\Sigma^* = \Sigma^*(J^*, \;^*_1, \;^*_2) \tag{4.17}$$

Moreover, in the examples considered here, attention is focused on the special case of a compressible Neo-Hookean material defined by

$$\rho_0^*\Sigma^* = \frac{1}{2}K(J^* \quad 1)^2 + \frac{1}{2}\mu(\;^*_1 \quad 3) \tag{4.18}$$

where $\{K, \mu\}$ are the small deformation bulk and shear modulus, respectively, and Poisson's ratio ν is defined such that

$$K = \frac{2\mu(1+\nu)}{3(1 \quad 2\nu)} \tag{4.19}$$

Unless otherwise stated, for the example problems considered in the later sections the material is taken to be compressible with the strain energy function (4.18) and with the material constants specified by

$$K = 1\,GPa \quad , \quad \mu = 0.6\,GPa \quad , \quad \nu = 0.25 \tag{4.20}$$

For the special examples which consider a nearly incompressible the material constants are specified by

$$K = 1000\,GPa \quad , \quad \mu = 0.6\,GPa \quad , \quad \nu = 0.4997 \tag{4.21}$$

5 Bubnov-Galerkin equations for a 3-D brick element

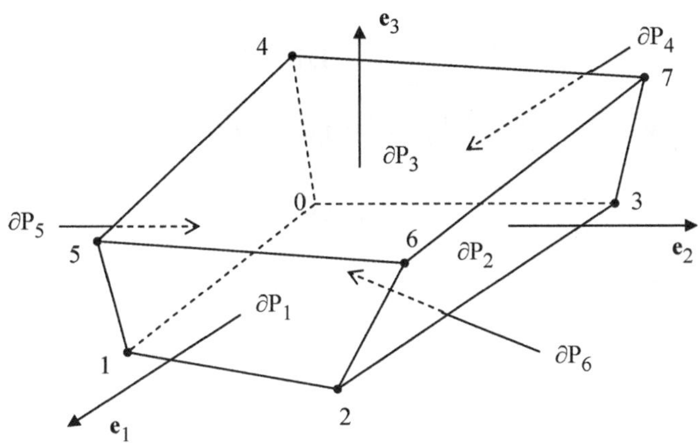

Figure 5.1. Sketch of a general brick CPE showing the numbering of the nodes and the surfaces.

Figure 5.1 shows a sketch of a 3-D brick element which occupies the region P with closed boundary ∂P characterized by the union of the six surfaces $\partial P_I\,(I = 1, 2, ..., 6)$. Within the context of the Bubnov-Galerkin approach based on tri-linear shape functions, the position vector $\mathbf{X}^*$ of a material point in the reference configuration is represented by

$$\mathbf{X}^* = \mathbf{X}^*(\theta^i) = \sum_{m=0}^{7} N^m(\theta^i)\mathbf{D}_m \tag{5.1}$$

where the shape functions $N^m(\theta^i)$ depend only on the convected coordinates θ^i and are given by

$$\begin{aligned} &N^0 = 1 \quad , \quad N^1 = \theta^1 \quad , \quad N^2 = \theta^2 \quad , \quad N^3 = \theta^3 \quad , \\ &N^4 = \theta^1\theta^2 \quad , \quad N^5 = \theta^1\theta^3 \quad , \quad N^6 = \theta^2\theta^3 \quad , \quad N^7 = \theta^1\theta^2\theta^3 \end{aligned} \tag{5.2}$$

and the reference element director vectors $\mathbf{D}_i$ (i=0,1,...,7) are constant vectors with $\mathbf{D}_i$ (i=1,2,3) being linearly independent

$$D^{1/2} = \mathbf{D}_1 \times \mathbf{D}_2 \bullet \mathbf{D}_3 > 0 \tag{5.3}$$

The locations of the nodes in the reference configuration are characterized by the constant reference nodal director vectors $\mathbf{D}_i$ $(i = 0, 1, ..., 7)$. In particular, the convected coordinates are limited by the lengths H_i $(i = 1, 2, 3)$, such that

$$|\theta_1| \leq \frac{H_1}{2} \quad , \quad |\theta_2| \leq \frac{H_2}{2} \quad , \quad |\theta_3| \leq \frac{H_3}{2} \tag{5.4}$$

and that

$$\begin{aligned}
\mathbf{D}_0 &= \mathbf{X}^*(\ \frac{H_1}{2}, \ \frac{H_2}{2}, \ \frac{H_3}{2}) \quad , \quad \mathbf{D}_1 = \mathbf{X}^*(\frac{H_1}{2}, \ \frac{H_2}{2}, \ \frac{H_3}{2}) \ , \\
\mathbf{D}_2 &= \mathbf{X}^*(\frac{H_1}{2}, \frac{H_2}{2}, \ \frac{H_3}{2}) \quad , \quad \mathbf{D}_3 = \mathbf{X}^*(\ \frac{H_1}{2}, \frac{H_2}{2}, \ \frac{H_3}{2}) \ , \\
\mathbf{D}_4 &= \mathbf{X}^*(\ \frac{H_1}{2}, \ \frac{H_2}{2}, \frac{H_3}{2}) \quad , \quad \mathbf{D}_5 = \mathbf{X}^*(\frac{H_1}{2}, \ \frac{H_2}{2}, \frac{H_3}{2}) \ , \\
\mathbf{D}_6 &= \mathbf{X}^*(\frac{H_1}{2}, \frac{H_2}{2}, \frac{H_3}{2}) \quad , \quad \mathbf{D}_7 = \mathbf{X}^*(\ \frac{H_1}{2}, \frac{H_2}{2}, \frac{H_3}{2})
\end{aligned} \tag{5.5}$$

Also, the lengths H_i are defined so that $\mathbf{D}_i$ $(i = 1, 2, 3)$ are unit vectors

$$|\mathbf{D}_1| = |\mathbf{D}_2| = |\mathbf{D}_3| = 1 \tag{5.6}$$

In the Bubnov-Galerkin approach it is also assumed that the position vector $\mathbf{x}^*$ of material points in the present configuration can be expressed using a representation of the form (5.1) with the reference element directors $\mathbf{D}_i$ replaced by the present element directors $\mathbf{d}_i(t)$, which are functions of time t only, such that

$$\mathbf{x}^* = \mathbf{x}^*(\theta^i, t) = \sum_{m=0}^{7} N^m(\theta^i)\mathbf{d}_m(t) \tag{5.7}$$

where it is assumed that $\mathbf{d}_i$ $(i = 1, 2, 3)$ are linearly independent vectors

$$d^{1/2} = \mathbf{d}_1 \times \mathbf{d}_2 \bullet \mathbf{d}_3 > 0 \tag{5.8}$$

In this regard, it should be noted that although the representation (5.1) is exact the expression (5.7) is an approximation of the deformation field in the element.

Now, the element directors can be expressed as functions of the nodal directors using a constant matrix A_{ij} $(i, j = 0, 1, ..., 7)$ that is determined by (5.1) and (5.5)

$$\mathbf{D}_i = \sum_{i=0}^{7} A_{ij}\mathbf{D}_j \quad , \quad \mathbf{d}_i = \sum_{i=0}^{7} A_{ij}\mathbf{d}_j \tag{5.9}$$

In this expression $\mathbf{d}_i(t)$ $(i = 0, 1, ..., 7)$ are the nodal director vectors that locate the present positions of the nodes of the element. Moreover, the element director velocities $\mathbf{w}_i$ and nodal director velocities $\bar{\mathbf{w}}_i$ are defined by

$$\mathbf{w}_i = \dot{\mathbf{d}}_i \quad , \quad \mathbf{w}_i = \dot{\mathbf{d}}_i \quad (i = 0, 1, ..., 7) \tag{5.10}$$

The objective of the Bubnov-Galerkin approach is to develop weak forms of equations for the director vectors $\mathbf{d}_i$ which represent an approximation of the partial differential equation (4.6a) expressing the balance of linear momentum. To this end, multiply (4.6b) by the weighting function $\phi(\theta^i)$ to deduce that

$$\phi m^* \dot{\mathbf{v}}^* = \phi m^* \mathbf{b}^* + \sum_{j=1}^{3} [(\phi \mathbf{t}^{*j})_{,j} \quad \mathbf{t}^{*j} \phi_{,j}] \tag{5.11}$$

Then, integrate this result over the region P to obtain the weak form

$$\frac{d}{dt} \int_P \phi \rho^* \mathbf{v}^* dv^* = \int_P \phi \rho^* \mathbf{b}^* dv^* + \int_{\partial P} \phi \mathbf{t}^* da^*$$
$$\sum_{j=1}^{3} \int_P g \quad {}^{1/2} \mathbf{t}^{*j} \phi_{,j} dv^* \tag{5.12}$$

where use has been made of the conservation of mass (4.6a) and the divergence theorem (4.5).

Next, it is convenient to introduce a number of quantities that are used in the CPE formulation. Specifically, the mass m of the element and the director inertia quantities y^{ij} are given by

$$m = \int_P \rho^* dv^* \quad ,$$
$$m y^{ij} = \int_P N^i N^j \rho^* dv^* = m y^{ji} \quad (i, j = 0, 1, ..., 7) \quad ,$$
$$y^{00} = 1 \tag{5.13}$$

the external assigned director couples $\mathbf{b}^i$ due to body forces and the director couples $\mathbf{m}^i$ due to surface tractions on the boundaries of the CPE are given by

$$m \mathbf{b}^i = \int_P N^i \rho^* \mathbf{b}^* dv^* \quad ,$$
$$\mathbf{m}^i = \int_{\partial P} N^i \mathbf{t}^* da^* \quad (i = 0, 1, ..., 7) \tag{5.14}$$

Also, the intrinsic director couples $\mathbf{t}^i$ are expressed by

$$\mathbf{t}^i = \sum_{j=1}^{3} \int_P g^{1/2} \mathbf{t}^{*j} N^i_{,j} dv^* \quad (i = 0, 1, ..., 7) \tag{5.15}$$

Then, using these definitions, the global balance laws (4.1- 4.3), the representation (5.7) and taking ϕ in (5.12) equal to N^i it is possible to derive the balance laws of the CPE. Specifically, the conservation of mass

$$\dot{m} = 0 \tag{5.16}$$

the balances of director momentum

$$\frac{d}{dt}(\sum_{j=0}^{7} my^{ij}\mathbf{w}_j) = m\mathbf{b}^i + \mathbf{m}^i \quad \mathbf{t}^i \;\; with \;\; \mathbf{t}^0 = \mathbf{0} \;,\; (i = 0, 1, .., 7) \tag{5.17}$$

and the balance of angular momentum

$$\frac{d}{dt}(\sum_{i=0}^{7}\sum_{j=0}^{7} \mathbf{d}_i \times my^{ij}\mathbf{w}_j) = \sum_{i=0}^{7} \mathbf{d}_i \times m\mathbf{b}^i + \sum_{i=0}^{7} \mathbf{d}_i \times \mathbf{m}^i \tag{5.18}$$

represent the balance laws of the CPE. In particular, it is noted that balances of director momentum (5.17) include the global form (4.2) of the balance of linear momentum for $i = 0$. Also, it can be shown that the director inertia coefficients y^{ij} are constants

$$\dot{y}^{ij} = 0 \tag{5.19}$$

Furthermore, using the representation (5.7) the rate of work W and kinetic energy K in (4.9) can be expressed in the forms

$$\begin{gathered} \mathsf{W} = \mathsf{W}_b + \mathsf{W}_c \;, \\ \mathsf{W}_b = \sum_{i=0}^{7} m\mathbf{b}^i \bullet \mathbf{w}_i \;,\quad \mathsf{W}_c = \sum_{i=0}^{7} \mathbf{m}^i \bullet \mathbf{w}_i \;, \\ \mathsf{K} = \sum_{i=0}^{7}\sum_{j=0}^{7} \frac{1}{2} my^{ij}\mathbf{w}_i \bullet \mathbf{w}_j \end{gathered} \tag{5.20}$$

where $\{\mathsf{W}_b, \mathsf{W}_c\}$ represent the rates of work done by body forces and surface tractions, respectively.

The main difference between the Bubnov-Galerkin approach and the CPE approach is the procedure for determining constitutive equations for the intrinsic director couples $\mathbf{t}^i$. This will be discussed in detail in the following sections.

6 Balance laws for a 3-D brick CPE (direct approach)

The nodes of the 3-D brick CPE shown in Figure 5.1 are characterized by the constant reference nodal directors $\mathbf{D}_i$ $(i = 0, 1, ..., 7)$ and by the present nodal director vectors $\mathbf{d}_i(t)$ $(i = 0, 1, ..., 7)$ which are functions of time. Then, the reference element directors $\mathbf{D}_i$ and present element directors $\mathbf{d}_i(t)$ are determined by the expressions (5.9) where A_{ij} is a constant matrix. Moreover, the element director velocities $\mathbf{w}_i$ and nodal director velocities $\mathbf{w}_i$ are given by (5.10).

In view of the restrictions (5.3) and (5.8) it is possible to define reciprocal vectors $\mathbf{D}^i$ and $\mathbf{d}^i$ $(i = 1, 2, 3)$ by formulas of the type (2.18) and (2.24) so that

$$\mathbf{D}_i \bullet \mathbf{D}^j = \delta_i^{\ j} \quad , \quad \mathbf{d}_i \bullet \mathbf{d}^j = \delta_i^{\ j} \tag{6.1}$$

Then, the kinematics of the CPE can be characterized by the deformation tensor $\mathbf{F}$ and its determinant J

$$\mathbf{F} = \mathbf{F}(t) = \sum_{i=1}^{3} \mathbf{d}_i \otimes \mathbf{D}^i \quad , \quad J = det(\mathbf{F}) = \frac{d^{1/2}}{D^{1/2}} > 0 \tag{6.2}$$

associated with homogeneous deformations and the vectors $\ _i$

$$\ _i = \mathbf{F}^{\ 1}\mathbf{d}_{i+3} \quad \mathbf{D}_{i+3} \quad , \quad \mathbf{d}_{i+3} = \mathbf{F}(\ _i + \mathbf{D}_{i+3}) \quad , \quad (i = 1, 2, 3, 4) \tag{6.3}$$

associated with inhomogeneous deformations. Furthermore, the rate of deformation tensor $\mathbf{L}$ and its symmetric part $\mathbf{D}$ are defined by

$$\mathbf{L} = \dot{\mathbf{F}}\mathbf{F}^{\ 1} = \sum_{i=1}^{3} \mathbf{w}_i \otimes \mathbf{d}^i \quad , \quad \mathbf{D} = \frac{1}{2}(\mathbf{L} + \mathbf{L}^T) = \mathbf{D}^T \tag{6.4}$$

so that

$$\mathbf{w}_i = \mathbf{L}\mathbf{d}_i \quad (i = 1, 2, 3) \quad ,$$
$$\mathbf{w}_{i+3} = \mathbf{L}\mathbf{d}_{i+3} + \mathbf{F}\dot{\ }_i \ , \quad \dot{\ }_i = \mathbf{F}^{\ 1}(\mathbf{w}_{i+3} \quad \mathbf{L}\mathbf{d}_{i+3}) \quad (i = 1, 2, 3, 4) \tag{6.5}$$

Within the context of the direct approach, the balance laws of the CPE are proposed as the conservation of mass (5.16), the balances of director momentum (5.17) and the balance of angular momentum (5.18). Also, the director inertia coefficients y^{ij} are constants (5.19) and the expressions for the rate of work W done on the CPE and its kinetic energy K are given by (2.20). Moreover, the rate of material dissipation   is proposed by an equation like (4.10), such that

$$d^{1/2} \quad = \mathsf{W} \quad \dot{\mathsf{K}} \quad m\dot{\Sigma} \geq 0 \tag{6.6}$$

where Σ is the strain energy function per unit mass m of the CPE.

Using the symmetry of the director inertia coefficients y^{ij}, the balances of director momentum (5.17) and introducing the tensor

$$d^{1/2}\mathbf{T} = \sum_{i=1}^{7} \mathbf{t}^i \otimes \mathbf{d}_i \tag{6.7}$$

it can be shown that the reduced form of the balance of angular momentum (5.18) requires $\mathbf{T}$ to be a symmetric tensor

$$\mathbf{T}^T = \mathbf{T} \tag{6.8}$$

which is similar to the restriction (4.8) associated with the three-dimensional theory.

Moreover, using the balances of director momentum (5.17), the rate of material dissipation can be expressed in the form

$$d^{1/2} \quad = \sum_{i=1}^{7} \mathbf{t}^i \bullet \mathbf{w}_i \quad m\dot{\Sigma} \geq 0 \tag{6.9}$$

However, with the help of (6.5) the mechanical power can be rewritten in the form

$$\sum_{i=1}^{7} \mathbf{t}^i \bullet \mathbf{w}_i = d^{1/2}\mathbf{T} \bullet \mathbf{D} + \sum_{i=1}^{4} \mathbf{F}^T\mathbf{t}^{\;i+3)} \bullet \;\dot{}_i \tag{6.10}$$

so the rate of material dissipation reduces to

$$d^{1/2} \quad = d^{1/2}\mathbf{T} \bullet \mathbf{D} + \sum_{i=1}^{4} \mathbf{F}^T\mathbf{t}^{\;i+3)} \bullet \;\dot{}_i \quad m\dot{\Sigma} \geq 0 \tag{6.11}$$

Now, comparison of (6.11) with the three-dimensional equation (4.11) suggests that $d^{1/2}\mathbf{T}$ is similar to the Cauchy stress. In fact, using the approximation (5.7) it follows that

$$\mathbf{g}_j = \sum_{i=1}^{7} N^i_{\;j}\mathbf{d}_i \quad (j = 1, 2, 3) \tag{6.12}$$

Thus, with the help of (2.22), (4.7), (5.15) and (6.7) it can be shown that $d^{1/2}\mathbf{T}$ is related to the volume integral of Cauchy stress

$$
\begin{aligned}
d^{1/2}\mathbf{T} &= \sum_{i=1}^{7}\sum_{j=1}^{3}\int_P g \quad {}^{1/2}\mathbf{t}^{*j} N^i_{\ ;j} dv^* \otimes \mathbf{d}_i \\
&= \int_P \mathbf{T}^* \Big(\sum_{j=1}^{3} \mathbf{g}^j \otimes \mathbf{g}_j \Big) dv^* \\
&= \int_P \mathbf{T}^* dv^*
\end{aligned}
\tag{6.13}
$$

Consequently, the volume averaged Cauchy stress $\mathbf{T}^*_{avg}$ is given by

$$
\mathbf{T}^*_{avg} = \frac{1}{v^*} d^{1/2}\mathbf{T} \tag{6.14}
$$

where with the help of (2.17), (3.5) and (5.7) the current volume of the element is given by

$$
\begin{aligned}
v^* &= \int_P dv^* \\
&= H_1 H_2 H_3 \Big[d^{1/2} + \frac{H_1^2}{12}\mathbf{d}_4 \times \mathbf{d}_5 \bullet \mathbf{d}_1 + \frac{H_2^2}{12}\mathbf{d}_6 \times \mathbf{d}_4 \bullet \mathbf{d}_2 \\
&+ \frac{H_3^2}{12}\mathbf{d}_5 \times \mathbf{d}_6 \bullet \mathbf{d}_3 \Big]
\end{aligned}
\tag{6.15}
$$

7 Constitutive equations for a hyperelastic CPE

The strain energy function Σ^* in (4.12) for a hyperelastic material is local in the sense that it characterizes the response of the material at a material point. In contrast, the 3-D brick CPE is a structure whose response depends on both the material and geometric properties and the structure. Consequently, the constitutive equations for the CPE necessarily combine material and geometric quantities.

A hyperelastic CPE is an ideal element in the same sense that a hyperelastic material is an ideal material. In particular, for a hyperelastic CPE it is assumed that the strain energy function Σ depends tacitly on the reference geometry of the CPE and explicitly on the deformations measures $\{\mathbf{C}, \ \ _i\}$

$$
\Sigma = \Sigma(\mathbf{C}, \ \ _i) \tag{7.1}
$$

Furthermore, the kinetic quantities $\{\mathbf{T}, \mathbf{t}^i\}$ are assumed to be independent of deformation rates and the rate of dissipation in (6.11) is assumed to vanish for all processes

$$d^{1/2} \quad = d^{1/2}\mathbf{T} \bullet \mathbf{D} + \sum_{i=1}^{4} \mathbf{F}^T \mathbf{t}^{\ i+3)} \bullet \ \dot{}_{i} \qquad m\dot{\Sigma} = 0 \tag{7.2}$$

Then, using the usual arguments it can be shown that the kinetic quantities are determined by derivatives of Σ

$$d^{1/2}\mathbf{T} = 2m\mathbf{F}\frac{\partial \Sigma}{\partial \mathbf{C}}\mathbf{F}^T \quad , \quad \mathbf{t}^{\ i+3)} = m\mathbf{F}^{\ T}\frac{\partial \Sigma}{\partial \ _i} \ (i = 1, 2, 3, 4) \tag{7.3}$$

with the remaining $\mathbf{t}^i$ $(i = 0, 1, 2, 3)$ being determined by (5.17) and (6.7)

$$\mathbf{t}^0 = 0 \quad , \quad \mathbf{t}^i = \left[d^{1/2}\mathbf{T} \quad \sum_{j=4}^{7} \mathbf{t}^j \otimes \mathbf{d}_j \right] \bullet \mathbf{d}^i \ (i = 1, 2, 3) \tag{7.4}$$

8 A nonlinear patch test

Following previous research on shells (Naghdi and Rubin (1995)), rods (Rubin (1996)) and points (Rubin (2000); Rubin (2001); Nadler and Rubin (2003)) it is possible to impose restrictions on the strain energy function Σ which ensure that the CPE produces solutions that are consistent with the exact three-dimensional theory for all homogeneous deformations of an arbitrary uniform homogeneous anisotropic elastic material. These restrictions are equivalent to a nonlinear patch test on the brick element. Specifically, confining attention to such a material it can be shown that the mass m is given by

$$m = \rho_0^* V^* \tag{8.1}$$

where the volume V^* of the CPE in its reference configuration is determined by using the representation (5.1) to deduce that

$$\begin{aligned} V^* = D^{1/2}V &= \int_P dV^* \\ &= H_1 H_2 H_3 \left[D^{1/2} + \frac{H_1^2}{12}\mathbf{D}_4 \times \mathbf{D}_5 \bullet \mathbf{D}_1 + \frac{H_2^2}{12}\mathbf{D}_6 \times \mathbf{D}_4 \bullet \mathbf{D}_2 \right. \\ &\left. + \frac{H_3^2}{12}\mathbf{D}_5 \times \mathbf{D}_6 \bullet \mathbf{D}_3 \right] \end{aligned} \tag{8.2}$$

and the quantity V has been introduced for convenience.

Now, using (3.3) and the kinematic assumption (5.7) it can be shown that

$$\begin{aligned}
\mathbf{F}^* &= \sum_{m=0}^{3} \left[\mathbf{F}\mathbf{D}_i + \sum_{i=1}^{4} N_{\ m}^{i+3)} \mathbf{F} \ \ \mathbf{D}_{i+3} + \ \ {}_i) \right] \otimes \mathbf{G}^m \\
&= \mathbf{F} \sum_{m=0}^{3} \left[\mathbf{G}_m \otimes \mathbf{G}^m + \sum_{i=1}^{4} N_{\ m}^{i+3)} \ \ {}_i \otimes \mathbf{G}^m \right] \\
&= \mathbf{F} \left[\mathbf{I} + \sum_{i=1}^{4} \sum_{m=0}^{3} N_{\ m}^{i+3)} \ \ {}_i \otimes \mathbf{G}^m \right]
\end{aligned} \tag{8.3}$$

Thus, the deformation will be three-dimensionally homogeneous with $\mathbf{F}^*$ being independent of the coordinates θ^i if $\ {}_i$ vanish

$$\mathbf{F}^* = \mathbf{F}(t) \quad for \quad \ {}_i = \mathbf{0} \ \ (i = 1, 2, 3, 4) \tag{8.4}$$

This result demonstrates that $\ {}_i$ are measures of inhomogeneous deformation. Moreover, Nadler and Rubin (2003) introduced the auxiliary deformation measure **F**

$$\mathbf{F} = \mathbf{F} \left(\mathbf{I} + \sum_{i=1}^{4} \ \ {}_i \otimes \mathbf{V}^i \right) \tag{8.5}$$

where the vectors $\mathbf{V}^i$ are defined by the reference geometry of the CPE such that

$$D^{1/2} V \mathbf{V}^i = \sum_{m=1}^{3} \int_P \left[N_{\ m}^{i+3)} \mathbf{G}^m \right] dV^* \ \ (i = 1, 2, 3, 4) \tag{8.6}$$

Consequently, with the help of (5.1), $\mathbf{V}^i$ are given by

$$\begin{aligned}
D^{1/2} V \mathbf{V}^1 &= H_1 H_2 H_3 \left[\frac{H_1^2}{12} \mathbf{D}_5 \times \mathbf{D}_1 + \frac{H_2^2}{12} \mathbf{D}_2 \times \mathbf{D}_6 \right] , \\
D^{1/2} V \mathbf{V}^2 &= H_1 H_2 H_3 \left[\frac{H_1^2}{12} \mathbf{D}_1 \times \mathbf{D}_4 + \frac{H_3^2}{12} \mathbf{D}_6 \times \mathbf{D}_3 \right] , \\
D^{1/2} V \mathbf{V}^3 &= H_1 H_2 H_3 \left[\frac{H_2^2}{12} \mathbf{D}_4 \times \mathbf{D}_2 + \frac{H_3^2}{12} \mathbf{D}_3 \times \mathbf{D}_5 \right] , \\
D^{1/2} V \mathbf{V}^4 &= \mathbf{0}
\end{aligned} \tag{8.7}$$

Also, integration of (8.3) over the region P_0 yields the result that $\mathbf{F}$ is the volume averaged three-dimensional deformation gradient (Loehnert et al. (2005))

$$\mathbf{F} = \frac{1}{D^{1/2}V} \int_P \mathbf{F}^* dV^* \tag{8.8}$$

Furthermore, for homogeneous deformations (8.4) of a hyperelastic material the Cauchy stress $\mathbf{T}^*$ is also independent of the coordinates. Consequently, with the help of (3.4), (4.6a), (4.13), (6.15), (8.1) and (8.2) it can be shown that for homogeneous deformations (6.13) requires

$$\begin{aligned} d^{1/2}\mathbf{T} = v^*\mathbf{T}^* &= D^{1/2}VJ^*\mathbf{T}^* \\ &= 2m\mathbf{F}^* \frac{\partial \Sigma^*}{\partial \mathbf{C}^*} \mathbf{F}^{*T} \\ &= 2m\mathbf{F} \frac{\partial \Sigma^*(\mathbf{C})}{\partial \mathbf{C}} \mathbf{F}^T \end{aligned} \tag{8.9}$$

Also, for homogeneous deformations (5.15) yields

$$\begin{aligned} \mathbf{t}^{\;\; i+3)} &= \sum_{m=1}^{3} \int_P \mathbf{T}^* \mathbf{g}^m N_{\;m}^{\;\; i+3)} dv^* \\ &= J\mathbf{T}^*\mathbf{F}^{\;\; T} \int_P \sum_{m=1}^{3} N_{\;m}^{\;\; i+3)} \mathbf{G}^m dV^* \\ &= D^{1/2}VJ\mathbf{T}^*\mathbf{F}^{\;\; T}\mathbf{V}^i \\ &= d^{1/2}\mathbf{T}\mathbf{F}^{\;\; T}\mathbf{V}^i \;\; (i = 1,2,3,4) \end{aligned} \tag{8.10}$$

Now, comparison of the results (8.9) and (8.10) with the constitutive equations (7.3) indicates that the CPE will satisfy the patch test provided that the strain energy function satisfies the restrictions that

$$\begin{aligned} \frac{\partial \Sigma}{\partial \mathbf{C}} &= \frac{\partial \Sigma^*(\mathbf{C})}{\partial \mathbf{C}} \;, \\ \frac{\partial \Sigma}{\partial \;\; _i} &= 2\mathbf{C}\frac{\partial \Sigma^*(\mathbf{C})}{\partial \mathbf{C}} \mathbf{V}^i \;\; (i = 1,2,3,4) \quad for \quad \;\; _i = \mathbf{0} \end{aligned} \tag{8.11}$$

Motivated by the work in (Nadler and Rubin (2003)) these restrictions can be simplified by writing the general form of the strain energy of the CPE as

$$\Sigma(\mathbf{C}, \;\; _i) = \Sigma^*(\mathbf{C}) + \Psi(\mathbf{C}, \;\; _i) \;, \quad (i = 1,2,3,4) \tag{8.12}$$

where here Ψ is taken to be a function of $\{\mathbf{C}, \ \ _i\}$ instead of $\{\mathbf{C}, \ \ _i\}$ as in (Nadler and Rubin (2003)). Next, taking the material derivative of (8.5) yields the results that

$$\dot{\mathbf{F}} = \mathbf{LF} + \mathbf{F}\Big(\sum_{i=1}^{4} \dot{\ }_i \otimes \mathbf{V}^i\Big) \ ,$$
$$\dot{\mathbf{C}} = \mathbf{F}^T\mathbf{F} \ \ ^T\dot{\mathbf{C}}\mathbf{F} \ \ ^1\mathbf{F} + \mathbf{F}^T\mathbf{F}\Big(\sum_{i=1}^{4} \dot{\ }_i \otimes \mathbf{V}^i\Big)$$
$$+ \Big(\sum_{i=1}^{4} \mathbf{V}^i \otimes \dot{\ }_i\Big)\mathbf{F}^T\mathbf{F} \tag{8.13}$$

so that using the chain rule of differentiation it can be shown that

$$\frac{\partial\Sigma}{\partial\mathbf{C}} = \mathbf{F} \ \ ^1\mathbf{F}\left[\frac{\partial\Sigma^*}{\partial\mathbf{C}} + \frac{\partial\Psi}{\partial\mathbf{C}}\right]\mathbf{F}^T\mathbf{F} \ \ ^T \ ,$$
$$\frac{\partial\Sigma}{\partial \ \ _i} = \frac{\partial\Psi}{\partial \ \ _i} + 2\mathbf{F}^T\mathbf{F}\left[\frac{\partial\Sigma^*}{\partial\mathbf{C}} + \frac{\partial\Psi}{\partial\mathbf{C}}\right]\mathbf{V}^i \ (i = 1, 2, 3, 4) \tag{8.14}$$

Thus, with the help of the representation (8.12) the restrictions (8.11) reduce to restrictions on only the inhomogeneous part of the strain energy

$$\frac{\partial\Psi(\mathbf{C}, \ \ _i)}{\partial\mathbf{C}} = \mathbf{0} \quad for \quad \ _i = \mathbf{0} \ (i = 1, 2, 3, 4) \ ,$$
$$\frac{\partial\Psi(\mathbf{C}, \ \ _i)}{\partial \ \ _m} = \mathbf{0} \quad for \quad \ _i = \mathbf{0} \ (i, m = 1, 2, 3, 4) \tag{8.15}$$

For general anisotropic materials it is not known how to propose a functional form for Ψ which includes dependence on the reference geometry that causes the CPE to produce accurate results for general irregular shaped elements experiencing bending dominated loads. However, progress made for isotropic materials will be discussed in the next sections.

9 A speci c form of the strain energy function for inhomogeneous

Using the definitions of the inhomogeneous strain measures κ_j^i in (Nadler and Rubin (2003); Jabareen and Rubin (2008a))

$$
\begin{aligned}
\kappa_1^1 &= H_2 \;\; {}_1 \bullet \mathbf{D}^1 \quad , \; \kappa_1^2 = H_1 \;\; {}_1 \bullet \mathbf{D}^2 \quad , \; \kappa_1^3 = H_3 \;\; {}_1 \bullet \mathbf{D}^3 \; , \\
\kappa_2^1 &= H_3 \;\; {}_2 \bullet \mathbf{D}^1 \quad , \; \kappa_2^2 = H_2 \;\; {}_2 \bullet \mathbf{D}^2 \quad , \; \kappa_2^3 = H_1 \;\; {}_2 \bullet \mathbf{D}^3 \; , \\
\kappa_3^1 &= H_1 \;\; {}_3 \bullet \mathbf{D}^1 \quad , \; \kappa_3^2 = H_3 \;\; {}_3 \bullet \mathbf{D}^2 \quad , \; \kappa_3^3 = H_2 \;\; {}_3 \bullet \mathbf{D}^3 \; , \\
\kappa_4^1 &= H_2 H_3 \;\; {}_4 \bullet \mathbf{D}^1 \; , \; \kappa_4^2 = H_1 H_3 \;\; {}_4 \bullet \mathbf{D}^2 \; , \; \kappa_4^3 = H_1 H_2 \;\; {}_4 \bullet \mathbf{D}^3
\end{aligned}
\tag{9.1}
$$

and the alternative variables b_i $(i = 1, 2, ..., 12)$ defined by

$$
b_i = \left\{ \kappa_1^1, \kappa_3^3, \kappa_1^2, \kappa_2^3, \kappa_2^1, \kappa_3^2, \kappa_1^3, \kappa_2^2, \kappa_3^1, \kappa_4^1, \kappa_4^2, \kappa_4^3 \right\}
\tag{9.2}
$$

it is convenient to replace the dependence of Ψ on $\;_j$ $(j = 1, 2, 3, 4)$ with dependence on b_j $(j = 1, 2, ..., 12)$ and express the strain energy function in the form

$$
\Sigma(\mathbf{C}, \;\; _j) = \Sigma^*(\mathbf{C}) + \Psi(\mathbf{C}, b_i) \quad (i = 1, 2, ..., 12)
\tag{9.3}
$$

Then, the constitutive equations for a hyperelastic CPE become

$$
\begin{aligned}
d^{1/2}\mathbf{T} &= 2m\mathbf{F}\left[\frac{\partial \Sigma^*}{\partial \mathbf{C}} + \frac{\partial \Psi}{\partial \mathbf{C}}\right]\mathbf{F}^T \; , \\
\mathbf{t}^4 &= \left[m\frac{\partial \Psi}{\partial b_1} H_2 \mathbf{d}^1 + m\frac{\partial \Psi}{\partial b_3} H_1 \mathbf{d}^2 + m\frac{\partial \Psi}{\partial b_7} H_3 \mathbf{d}^3\right] \\
&\quad + d^{1/2}\mathbf{T} \;\; \mathbf{F} \;\; ^T\mathbf{V}^1\Big) \; , \\
\mathbf{t}^5 &= \left[m\frac{\partial \Psi}{\partial b_5} H_3 \mathbf{d}^1 + m\frac{\partial \Psi}{\partial b_8} H_2 \mathbf{d}^2 + m\frac{\partial \Psi}{\partial b_4} H_1 \mathbf{d}^3\right] \\
&\quad + d^{1/2}\mathbf{T} \;\; \mathbf{F} \;\; ^T\mathbf{V}^2\Big) \; , \\
\mathbf{t}^6 &= \left[m\frac{\partial \Psi}{\partial b_9} H_1 \mathbf{d}^1 + m\frac{\partial \Psi}{\partial b_6} H_3 \mathbf{d}^2 + m\frac{\partial \Psi}{\partial b_2} H_2 \mathbf{d}^3\right] \\
&\quad + d^{1/2}\mathbf{T} \;\; \mathbf{F} \;\; ^T\mathbf{V}^3\Big) \; , \\
\mathbf{t}^7 &= \left[m\frac{\partial \Psi}{\partial b_{10}} H_2 H_3 \mathbf{d}^1 + m\frac{\partial \Psi}{\partial b_{11}} H_1 H_3 \mathbf{d}^2 + m\frac{\partial \Psi}{\partial b_{12}} H_1 H_2 \mathbf{d}^3\right]
\end{aligned}
\tag{9.4}
$$

with the remaining expressions for $\mathbf{t}^i\,(i = 0, 1, 2, 3)$ are given by (7.4). Moreover, the special case considered in (Jabareen and Rubin (2008a)) takes to be a quadratic function of b_i which is independent of $\mathbf{C}$, such that

$$2m\Psi = \frac{D^{1/2}V\mu}{6(1 \quad \nu)}\left[\sum_{i=1}^{12}\sum_{j=1}^{12} B_{ij}b_ib_j\right] \tag{9.5}$$

where $\{\mu, \nu\}$ are the shear modulus and Poisson's ratio associated with the small deformation response and B_{ij} is a symmetric matrix. As a special case, higher-order hourglass modes are uncoupled from bending and torsional modes so that (Nadler and Rubin (2003))

$$\begin{aligned}
B_{ij} &= 0 \quad for \quad i = 10 \quad and \quad j \neq 10 \;, \\
B_{ij} &= 0 \quad for \quad i = 11 \quad and \quad j \neq 11 \;, \\
B_{ij} &= 0 \quad for \quad i = 12 \quad and \quad j \neq 12 \;, \\
B_{\;10\;10)} &= \frac{(1 \quad \nu)}{24}\left[\frac{2(3 \quad \nu)}{(3 \quad 2\nu)} + \frac{H_1^2}{H_2^2} + \frac{H_1^2}{H_3^2}\right] \;, \\
B_{\;11\;11)} &= \frac{(1 \quad \nu)}{24}\left[\frac{2(3 \quad \nu)}{(3 \quad 2\nu)} + \frac{H_2^2}{H_1^2} + \frac{H_2^2}{H_3^2}\right] \;, \\
B_{\;12\;12)} &= \frac{(1 \quad \nu)}{24}\left[\frac{2(3 \quad \nu)}{(3 \quad 2\nu)} + \frac{H_3^2}{H_1^2} + \frac{H_3^2}{H_2^2}\right]
\end{aligned} \tag{9.6}$$

The remaining 45 values of B_{ij} need to be determine by matching exact solutions to specific problems.

10 Determination of the constitutive coefficients

The coefficients B_{ij} of the strain energy Ψ of inhomogeneous deformations (9.5) can be determined by matching exact solutions of the linearized theory of an isotropic elastic material. Specifically, these coefficients were determined in (Nadler and Rubin (2003)) by matching exact solutions of pure bending and pure torsion of a rectangular parallelepiped. Then, the same functional forms of B_{ij} were used for elements with general reference shapes. In (Loehnert et al. (2005)) it was shown that the resulting CPE exhibited robust, accurate response to a number of problems which typically exhibit unphysical locking or hourglassing in other element formulations. However, it was also shown there that the CPE exhibited undesirable sensitivity to irregularity of the reference element shape.

Recently, Boerner et al. (2007) have proposed a numerical method for determining coefficients in a quadratic form of the strain energy function for inhomogeneous deformations of a 2-D plane strain formulation of the CPE. This numerical approach produces improved response for irregular shaped elements. Jabareen and Rubin (2007b) developed analytical forms for B_{ij} which cause an improved 3-D brick CPE to yield results that are relatively insensitive to element irregularity for a number of problems. However, it was observed that the accuracy of this improved CPE for out-of-plane bending of a rhombic plate degrades as the angle of the plate decreases from 90°. This is because the coefficients developed in (Jabareen and Rubin (2007b)) were not based on out-of-plane bending solutions. Later, Jabareen and Rubin (2008a) developed a generalized CPE which removed the deficiency in the improved CPE. Specifically, the functional forms for B_{ij} were generalized to include full coupling of bending and torsional modes of deformation. Also, the generalized CPE was obtained by considering out-of-plane bending solutions in addition to in-plane bending solutions.

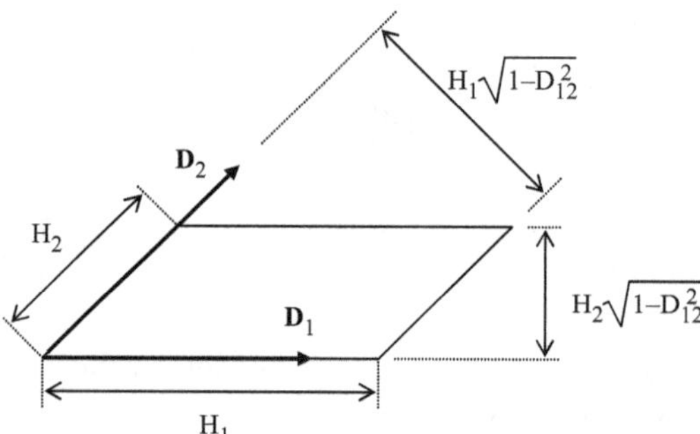

Figure 10.1. Sketch of the cross-section of the parallelepiped element E1.

It was observed in (Jabareen and Rubin (2007b) and Jabareen and Rubin (2008a)) that in order to develop functional forms for B_{ij} which produce a CPE that is relatively insensitive to element irregularity it is sufficient to focus attention on the bending and torsion response of elements which are parallelepipeds with two right angles. Specifically, with reference to the base vectors $\mathbf{e}_i$ of a fixed rectangular Cartesian coordinate system it is convenient to introduce the metric D_{ij}

$$D_{ij} = \mathbf{D}_i \bullet \mathbf{D}_j \tag{10.1}$$

and consider the three elements E1-E3 defined by (see Figure 10.1 for E1)

Element E1 $\quad (D_{12} \neq 0\,,\; D_{13} = 0\,,\; D_{23} = 0)$

$$\mathbf{D}_1 = \mathbf{e}_1 \;,\;\; \mathbf{D}_2 = D_{12}\mathbf{e}_1 + \sqrt{1 \quad D_{12}^2}\mathbf{e}_2 \;,\;\; \mathbf{D}_3 = \mathbf{e}_3 \;,$$

$$\mathbf{D}_i = 0 \;\; (i = 0, 4, 5, 6, 7) \tag{10.2a}$$

Element E2 $\quad (D_{12} = 0\,,\; D_{13} \neq 0\,,\; D_{23} = 0)$

$$\mathbf{D}_1 = \mathbf{e}_1 \;,\;\; \mathbf{D}_2 = \mathbf{e}_2 \;,\;\; \mathbf{D}_3 = D_{13}\mathbf{e}_1 + \sqrt{1 \quad D_{13}^2}\mathbf{e}_3 \;,$$

$$\mathbf{D}_i = 0 \;\; (i = 0, 4, 5, 6, 7) \tag{10.2b}$$

Element E3 $\quad (D_{12} = 0\,,\; D_{13} = 0\,,\; D_{23} \neq 0)$

$$\mathbf{D}_1 = \mathbf{e}_1 \;,\;\; \mathbf{D}_2 = \mathbf{e}_2 \;,\;\; \mathbf{D}_3 = D_{23}\mathbf{e}_2 + \sqrt{1 \quad D_{23}^2}\mathbf{e}_3 \;,$$

$$\mathbf{D}_i = 0 \;\; (i = 0, 4, 5, 6, 7) \tag{10.2c}$$

It then follows from (8.2) and (8.7) that for these elements

$$V = H_1 H_2 H_3 \;\;,\;\; \mathbf{V}^i = \mathbf{0} \;\; (i = 1, 2, 3, 4) \tag{10.3}$$

and the position vector $\mathbf{X}^*$ in (5.1) reduces to

$$\mathbf{X}^* = \sum_{j=1}^{3} \theta^j \mathbf{D}_j \tag{10.4}$$

where $\mathbf{D}_0$ has been set equal to zero. Also, for the example problems considered in this section use is made of linearized constitutive equations associated with the compressible Neo-Hookean strain energy function (4.18).

Now, with reference to the base vectors $\mathbf{e}_i'$ of another fixed rectangular Cartesian coordinate system, the components $\{X_i^{*\prime},\, u_i^{*\prime},\, T_{ij}^{*\prime}\}$ of the position vector $\mathbf{X}^*$, the displacement vector $\mathbf{u}^*$ and the stress tensor $\mathbf{T}^*$, respectively, of an exact solution of the linear equations of elasticity can be expressed in the forms

$$\mathbf{X}^* = \sum_{i=1}^{3} X_i^{*\prime}\mathbf{e}_i' \;\;,\;\; \mathbf{u}^* = \sum_{i=1}^{3} u_i^{*\prime}\mathbf{e}_i' \;,$$

$$\mathbf{T}^* = \sum_{i=1}^{3}\sum_{j=1}^{3} T_{ij}^{*\prime}(\mathbf{e}_i' \otimes \mathbf{e}_j') \tag{10.5}$$

Moreover, the classical pure bending solution (e.g. Sokolnikoff (1956)) of the three dimensional equations of isotropic elasticity for a rectangular parallelepiped can be written in the form

$$\begin{aligned}\mathbf{u}^* =&\ \ \gamma X_1^{*\prime} X_2^{*\prime}\Big)\mathbf{e}_1^\prime \quad \frac{1}{2}\gamma\Big[\ \ X_1^{*\prime}\Big)^2 + \nu\ \ X_2^{*\prime}\Big)^2 \quad \nu\ \ X_3^{*\prime}\Big)^2\Big]\mathbf{e}_2^\prime \\ &\ \ \gamma\nu X_2^{*\prime} X_3^{*\prime}\Big)\mathbf{e}_3^\prime\ , \\ \mathbf{T}^* =&\ 2\mu(1+\nu^*)\gamma X_2^{*\prime}(\mathbf{e}_1^\prime \otimes \mathbf{e}_1^\prime)\end{aligned} \tag{10.6}$$

where γ controls the magnitude of the pure bending. Similarly, the simple torsion-like solution in (Jabareen and Rubin (2007c)) can be expressed in the form

$$\begin{aligned}\mathbf{u}^* =&\ \ \ \omega\phi X_2^{*\prime} X_3^{*\prime}\Big)\mathbf{e}_1^\prime \quad \ \omega X_1^{*\prime} X_3^{*\prime}\Big)\mathbf{e}_2^\prime + \ \ \omega X_1^{*\prime} X_2^{*\prime}\Big)\mathbf{e}_3^\prime\ , \\ \mathbf{T}^* =&\ \mu\omega\Big[\ \ \ (1+\phi)X_3^{*\prime}(\mathbf{e}_1^\prime \otimes \mathbf{e}_2^\prime + \mathbf{e}_2^\prime \otimes \mathbf{e}_1^\prime) \\ &+ (1\ \ \ \phi)X_2^{*\prime}(\mathbf{e}_1^\prime \otimes \mathbf{e}_3^\prime + \mathbf{e}_3^\prime \otimes \mathbf{e}_1^\prime)\Big]\end{aligned} \tag{10.7}$$

where the constant ω is the twist per unit length in the $\mathbf{e}_1^\prime$ direction and the constant ϕ controls the warping of the cross-section with unit normal $\mathbf{e}_1^\prime$. Moreover, with the help of (10.4) and (10.5) the components $X_i^{*\prime}$ are determined by the convected coordinates θ^j

$$X_i^{*\prime} = \sum_{j=1}^{3}(\mathbf{e}_i^\prime \bullet \mathbf{D}_j)\theta^j \tag{10.8}$$

so that the solutions (10.6) and (10.7) can be expressed as function of θ^j.

Within the context of the linear theory of a CPE (Nadler and Rubin (2003)) the director displacements $\boldsymbol{\delta}_i$ are defined such that

$$\mathbf{d}_i = \mathbf{D}_i + \boldsymbol{\delta}_i \ \ (i = 0, 1, ..., 7) \tag{10.9}$$

and for the special elements defined by (10.2) the linearized forms of the inhomogeneous strains $\ _i$ become

$$\ _i = \boldsymbol{\delta}_{i+3} \ \ (i = 1, 2, 3, 4) \tag{10.10}$$

As explained in (Nadler and Rubin (2003)), the values $\boldsymbol{\delta}_i^*$ of the element director displacements $\boldsymbol{\delta}_i$ which correspond to the exact displacement field $\mathbf{u}^*$ need to be properly defined. Specifically, for these element shapes the values $\boldsymbol{\delta}_i^*$ are determined by the equations in (Nadler and Rubin (2003)) which connect $\boldsymbol{\delta}_i^*$ to integrals over the reference element region P_0 of derivatives of $\mathbf{u}^*$ with respect to the convected coordinates

$$\boldsymbol{\delta}_0^* = \frac{1}{V^*}\int_P \mathbf{u}^* dV^* \quad , \quad \boldsymbol{\delta}_i^* = \frac{1}{V^*}\int_P \frac{\partial \mathbf{u}^*}{\partial \theta^i} dV^* \;\; (i=1,2,3) \;\; ,$$

$$\boldsymbol{\delta}_4^* = \frac{1}{V^*}\int_P \frac{\partial^2 \mathbf{u}^*}{\partial \theta^1 \partial \theta^2} dV^* \;\; , \;\; \boldsymbol{\delta}_5^* = \frac{1}{V^*}\int_P \frac{\partial^2 \mathbf{u}^*}{\partial \theta^1 \partial \theta^3} dV^* \;\; ,$$

$$\boldsymbol{\delta}_6^* = \frac{1}{V^*}\int_P \frac{\partial^2 \mathbf{u}^*}{\partial \theta^2 \partial \theta^3} dV^* \;\; , \;\; \boldsymbol{\delta}_7^* = \frac{1}{V^*}\int_P \frac{\partial^3 \mathbf{u}^*}{\partial \theta^1 \partial \theta^2 \partial \theta^3} dV^* \qquad (10.11)$$

In particular, for the exact solutions (10.6) and (10.7) and the element shapes (10.2) it can be shown that these expressions yield

$$\boldsymbol{\delta}_1^* = \boldsymbol{\delta}_2^* = \boldsymbol{\delta}_3^* = \boldsymbol{\delta}_7^* = \mathbf{0} \qquad (10.12)$$

so that when $\boldsymbol{\delta}_i$ are replaced by the exact values $\boldsymbol{\delta}_i^*$ the linearized values of κ_4^i vanish so that [(9.2)]

$$\kappa_4^1 = \kappa_4^2 = \kappa_4^3 = 0 \quad or \quad b_{10} = b_{11} = b_{12} = 0 \qquad (10.13)$$

and with the help of (7.4) and (9.5) the linearized forms of the constitutive equations (9.4) reduce to

$$d^{1/2}\mathbf{T} = \mathbf{0} \;\; , \;\; \mathbf{t}^i = \mathbf{0} \;\; (i = 0,1,2,3,7) \;\; ,$$

$$\mathbf{t}^4 = \frac{D^{1/2}V\mu}{6(1\;\;\nu)} \sum_{j=1}^{12} \Big[B_{1j}H_2\mathbf{D}^1 + B_{3j}H_1\mathbf{D}^2 + B_{7j}H_3\mathbf{D}^3\Big] b_j \;\; ,$$

$$\mathbf{t}^5 = \frac{D^{1/2}V\mu}{6(1\;\;\nu)} \sum_{j=1}^{12} \Big[B_{5j}H_3\mathbf{D}^1 + B_{8j}H_2\mathbf{D}^2 + B_{4j}H_1\mathbf{D}^3\Big] b_j \;\; ,$$

$$\mathbf{t}^6 = \frac{D^{1/2}V\mu}{6(1\;\;\nu)} \sum_{j=1}^{12} \Big[B_{9j}H_1\mathbf{D}^1 + B_{6j}H_3\mathbf{D}^2 + B_{2j}H_2\mathbf{D}^3\Big] b_j \qquad (10.14)$$

where $\mathbf{d}^i$ have been replaced by the reference values $\mathbf{D}^i$. Also, the values of $\mathbf{m}^i$ in (5.14) associated with the exact solutions (10.6) and (10.7) are given by

$$\mathbf{m}^i = \mathbf{0} \;\; (i = 0,1,2,3,7) \;\; ,$$

$$\mathbf{m}^i = \int_{\partial P} N^i \mathbf{T}^* \mathbf{N}^* dA^* \;\; (i = 4,5,6) \qquad (10.15)$$

where ∂P_0 is the reference boundary of the CPE, $\mathbf{N}^*$ is the unit outward normal to ∂P_0 and dA^* is the reference element of area. It then follows that within the context of the linearized theory, the equations of equilibrium associated with the bending (10.6) and torsion (10.7) solutions reduce to three vector equations

$$\mathbf{t}^i \quad \mathbf{m}^i = \mathbf{0} \;\; (i = 4, 5, 6) \tag{10.16}$$

Analytical expressions for B_{ij} can be developed by matching the solutions (10.6) and (10.7) for each of the element shapes (10.2). Specifically, with reference to the element shape E1 in (10.2a) consider six bending solutions associated with specifications of the orientations of $\mathbf{e}'_i$ relative to $\mathbf{D}_i$

$$\text{Bending B1:} \quad \mathbf{e}'_1 = \mathbf{D}_1 \ , \ \mathbf{e}'_3 = \mathbf{D}_3 \ , \tag{10.17a}$$
$$\text{Bending B2:} \quad \mathbf{e}'_1 = \mathbf{D}_1 \ , \ \mathbf{e}'_2 = \mathbf{D}_3 \ , \tag{10.17b}$$
$$\text{Bending B3:} \quad \mathbf{e}'_1 = \mathbf{D}_2 \ , \ \mathbf{e}'_3 = \mathbf{D}_3 \ , \tag{10.17c}$$
$$\text{Bending B4:} \quad \mathbf{e}'_1 = \mathbf{D}_2 \ , \ \mathbf{e}'_2 = \mathbf{D}_3 \ , \tag{10.17d}$$
$$\text{Bending B5:} \quad \mathbf{e}'_1 = \mathbf{D}_3 \ , \ \mathbf{e}'_3 = \mathbf{D}_2 \ , \tag{10.17e}$$
$$\text{Bending B6:} \quad \mathbf{e}'_1 = \mathbf{D}_3 \ , \ \mathbf{e}'_3 = \quad \mathbf{D}_1 \tag{10.17f}$$

Also, consider two torsion solutions associated with specifications

$$\text{Torsion T1:} \quad \mathbf{e}'_1 = \mathbf{D}_1 \ , \ \mathbf{e}'_3 = \mathbf{D}_3 \ , \tag{10.18a}$$
$$\text{Torsion T2:} \quad \mathbf{e}'_1 = \mathbf{D}_2 \ , \ \mathbf{e}'_3 = \mathbf{D}_3 \tag{10.18b}$$

For each bending and torsion solution the exact values $\boldsymbol{\delta}^*_i$ are determined by (10.11) the linearized values of b_i are determined using (9.1), (9.2) and (10.10) with $\boldsymbol{\delta}_i$ replaced by $\boldsymbol{\delta}^*_i$ and the resulting constitutive equations for $\mathbf{t}^i$ are determined by (10.14). Also, the values of the warping constant ϕ corresponding to nearly pure torsion being determined by

$$\text{Torsion T1:} \quad \mathbf{m}^6 \bullet \mathbf{D}_1 = 0 \;\Rightarrow\; \phi = \frac{H_2^2 \;\; 1 \quad D_{12}^2\big) \quad H_3^2}{H_2^2 \;\; 1 \quad D_{12}^2\big) + H_3^2} \ , \tag{10.19a}$$

$$\text{Torsion T2:} \quad \mathbf{m}^5 \bullet \mathbf{D}_2 = 0 \;\Rightarrow\; \phi = \frac{H_1^2 \;\; 1 \quad D_{12}^2\big) \quad H_3^2}{H_1^2 \;\; 1 \quad D_{12}^2\big) + H_3^2} \tag{10.19b}$$

For each bending solution the value of γ can be eliminated in the resulting equations of equilibrium (10.16) and the value of ω can be eliminated from each of the equations of equilibrium associated with the torsion solutions. Also, the values (10.19) are used in the resulting torsion equations. It therefore follows that the each of the solutions (B1)-(B6), (T1) and (T2) yield nine scalar equations of equilibrium which total 72 scalar equations to determine the values of B_{ij} as functions of H_i and D_{12}. Some of these scalar equations are trivially satisfied and others are redundant. In particular, using a symbolic program like *Maple* it can be shown that these equations can be solved for B_{ij} such that

$$det(B_{ij}) > 0 \tag{10.20}$$

Similar procedures can be used to define bending and torsion solutions for the element shapes E2 and E3 and the resulting equations can be solved for B_{ij} to determine the dependence on the metrics D_{13} and D_{23}. Next, introducing the auxiliary variables $\{\lambda_{12},\ \lambda_{13},\ \lambda_{23}\}$ defined by

$$\begin{aligned}
&For \quad D_{12}^2 + D_{13}^2 + D_{23}^2 = 0 : \\
&\qquad \lambda_{12} = \lambda_{13} = \lambda_{23} = 0 \ , && (10.21a) \\
&For \quad D_{12}^2 + D_{13}^2 + D_{23}^2 > 0 : \\
&\qquad \lambda_{12} = \frac{D_{12}^2}{D_{12}^2 + D_{13}^2 + D_{23}^2} \ , \quad \lambda_{13} = \frac{D_{13}^2}{D_{12}^2 + D_{13}^2 + D_{23}^2} \ , \\
&\qquad \lambda_{23} = \frac{D_{23}^2}{D_{12}^2 + D_{13}^2 + D_{23}^2} && (10.21b)
\end{aligned}$$

it is possible to denote the values of B_{ij} associated with the solutions of the three elements E1-E3 in (10.2) by B_{ij}^{12} for E1, by B_{ij}^{13} for E2, and by B_{ij}^{23} for E3. Also, the matrix B_{ij}^{0} is defined so that it yields a strain energy function Ψ equivalent to that obtained in (Nadler and Rubin (2003)) for a rectangular parallelepiped, when the value of the torsion function $b^*(1)$ is taken to be 1/2 as suggested in (Jabareen and Rubin (2007c)). Then, the general expression $B_{ij}(D_{12}, D_{13}, D_{23})$ which combines these solutions is given by

$$\begin{aligned}
B_{ij}(D_{12}, D_{13}, D_{23}) = (1 \quad \lambda_{12} \quad \lambda_{13} \quad \lambda_{23})B_{ij}^{0} \\
+ \lambda_{12}B_{ij}^{12} + \lambda_{13}B_{ij}^{13} + \lambda_{23}B_{ij}^{23}
\end{aligned} \tag{10.22}$$

Now, using the definitions (10.21) it follows that each of the coefficients $\{(1 \quad \lambda_{12} \quad \lambda_{13} \quad \lambda_{23}),\ \lambda_{12},\ \lambda_{13},\ \lambda_{23}\}$ is non-negative and that at least one of them is positive. Also, each of the matrices $\{B_{ij}^{0}, B_{ij}^{12}, B_{ij}^{13}, B_{ij}^{23}\}$ is positive definite so that the combined matrix $B_{ij}(D_{12}, D_{13}, D_{23})$ is also positive definite for all reference element shapes.

11 A test for path-dependence

The formulation of the CPE for nonlinear elasticity treats the element as a structure and determines the kinetic quantities by derivatives of a strain energy function so that the dissipation vanishes. It therefore follows that the CPE formulation is automatically hyperelastic and predicts path-independent results. In contrast, element formulations which modify full integration methods like those associated with enhanced strain or incompatible mode methods can introduce path-dependence of the results. Although an analytical proof is required to ensure that an element formulation is hyperelastic for all deformations, only a single calculation is needed to prove that an element formulation is path-dependent.

Jabareen and Rubin (2007a) introduced a simple simulation which can be used to test element formulations for path-dependence and the elements in Table 1.1 were tested. In the examples considered here and in the rest of the text the elements ABBAQUS-6, ADINA-2, ANSYS-3 and FEAP-3 in Table 1.1 are denoted by (AB), (AD), (F), respectively. Also, the generalized CPE is denoted by (C).

It has been shown in (Jabareen and Rubin (2007a)) that the response of (F) is similar to that of other enhanced strain and incompatible mode elements. Therefore, for most of the example problems presented in the following sections, comparisons will be limited to the element in FEAP. However, the results of the (Q1P0) 3-D brick element and the mixed higher order nine node quadrilateral element [denoted by (HO9)] in FEAP will be used for comparison of some of the examples using nearly incompressible material response.

In order to test potential path-dependency of element formulations, consider a single brick element which is a cube in its reference configuration with edges of length $L = 1\,m$ (see Figure 11.1). The four nodes located by $\mathbf{X}_i$ $(i = 1, 2, 3, 4)$ are fixed, the nodes $\mathbf{X}_i$ $(i = 5, 6, 7)$ are free and the node $\mathbf{X}_8$ is deformed to the location $\mathbf{x}_8$ by the displacement $\mathbf{u}_8$

$$\mathbf{x}_8 = \mathbf{X}_8 + \mathbf{u}_8 \tag{11.1}$$

Specifically, the displacement $\mathbf{u}_8$ is characterized by a sequence of straight line segments that connect the end points A-F shown in Figure 11.1 which are characterized by

$$
\begin{aligned}
&\mathbf{u}_A = \mathbf{0} \;,\quad \mathbf{u}_B = \quad \beta\mathbf{e}_1 \;,\quad \mathbf{u}_C = \quad \beta\mathbf{e}_1 \quad \beta\mathbf{e}_2 \;, \\
&\mathbf{u}_D = \quad \beta\mathbf{e}_1 \quad \beta\mathbf{e}_2 \quad \beta\mathbf{e}_3 \;,\quad \mathbf{u}_E = \quad \beta\mathbf{e}_2 \quad \beta\mathbf{e}_3 \;, \\
&\mathbf{u}_F = \quad \beta\mathbf{e}_3 \;,\quad \beta = 0.25\, m
\end{aligned}
\tag{11.2}
$$

The total external work done on the element is given by

$$
W = \int \mathbf{f}_8 \bullet \dot{\mathbf{u}}_8 \, dt \tag{11.3}
$$

where $\mathbf{f}_8$ is the external nodal force applied to node 8.

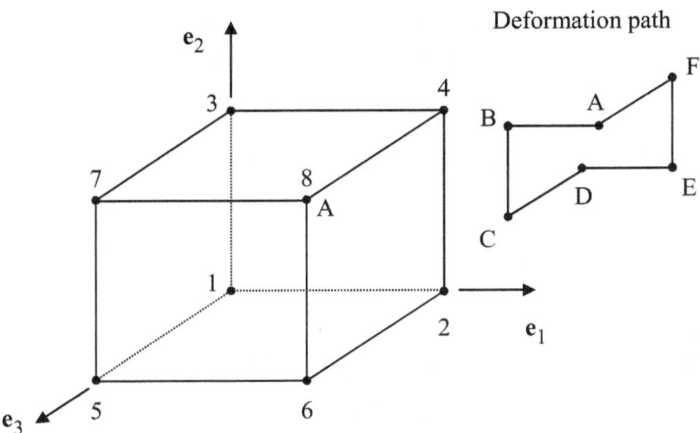

Figure 11.1. Sketch of the cubical element showing the numbering of the nodes and the paths used to test path-dependency of various element formulations.

To quantify the error associated with path-dependency it is convenient to introduce the quantities

$$
\begin{aligned}
&W_A^* = W_{ABCDEFA} \;,\quad W_A^{**} = W_{AFEDEFA} \;, \\
&W_D^* = W_{ABCD} \;,\quad W_D^{**} = W_{AFED} \;, \\
&E_A^* = \frac{W_A^*}{W_D^*} \;,\quad E_A^{**} = \frac{W_A^{**}}{W_D^*} \;,\quad E_A = \frac{W_D^* \quad W_D^{**}}{W_D^*}
\end{aligned}
\tag{11.4}
$$

Here, W_A^* denotes the work done on the element in a single cycle of deformation following the path $ABCDEFA$, W_A^{**} denotes the work done in a single cycle of the path $AFED$ and its reverse path $DEFA$, W_D^* is the work done during the path $ABCD$ and W_D^{**} is the work done during the path $AFED$ to the same point D. Also, E_A^* and E_A^{**} are the relative errors for the cycles associated with W_A^* and W_A^{**}, respectively and E_D is the relative error associated with the two different paths to the point D. The trapezoidal rule was used to integrate (11.3) and each segment of the deformation was divided into $N = 250$ equal steps to ensure accuracy.

Three types of element response are possible: hyperelastic, Cauchy elastic and hypoelastic. For hyperelastic element response the nodal forces maintaining equilibrium of any configuration and the work done between two configurations are both path-independent. For Cauchy elastic element response the nodal forces maintaining equilibrium of any configuration are path-independent but the work done between two configurations is path-dependent. For hypoelastic element response the nodal forces maintaining equilibrium of any configuration and the work done between two configurations are both path-dependent.

Table 11.1. Path-independence tests. Errors in the work and description of the type of elastic response.

Element	W_D (MJ)	(%)	(%)	$_D$ (%)	Type of elasticity
C	40.6	8.3E-5	2.1E-14	-8.3E-5	Hyper
AB	36.8	0.393	0.011	-3.18	HYPO
AD	38.9	1.1E-5	-1.0E-8	-1.1E-5	Hyper
AN	36.8	0.393	0.011	-3.18	HYPO
F	41.3	8.6E-5	-2.6E-15	-8.6E-5	Hyper

Table 11.1 presents the results for the Cosserat point element (C) and for the other enhanced strain/incompatible mode elements. The theoretical values of $\{E_A^*, E_A^{**}, E_D\}$ for the Cosserat point element are zero. Consequently, the numerical values for the Cosserat solution in Table 11.1 represent the combined numerical error due to: the convergence criterion used to satisfy equilibrium, machine precision and numerical integration of the work done using the trapezoidal rule. Thus, the error in the constitutive equation of a particular element can be determined by comparing the relative error with that of the Cosserat element. Furthermore, it is noted that the differences in the values of the work W_D^* given in Table 11.1 reflect differences in the specific treatment of inhomogeneous deformations in each of the elements. Moreover, the errors E_A^{**} associated with a cycle composed of a path and its reverse path are typically smaller than those E_A^* associated with a gen-

eral cycle. The errors E_D associated with two different paths to the same point can be up to 10 times those of E_A^*. Also, it is noted that negative values of E_A^* or E_A^{**} in Table 11.1 indicate that the element generates energy whereas positive values of these quantities indicate that the element dissipates energy.

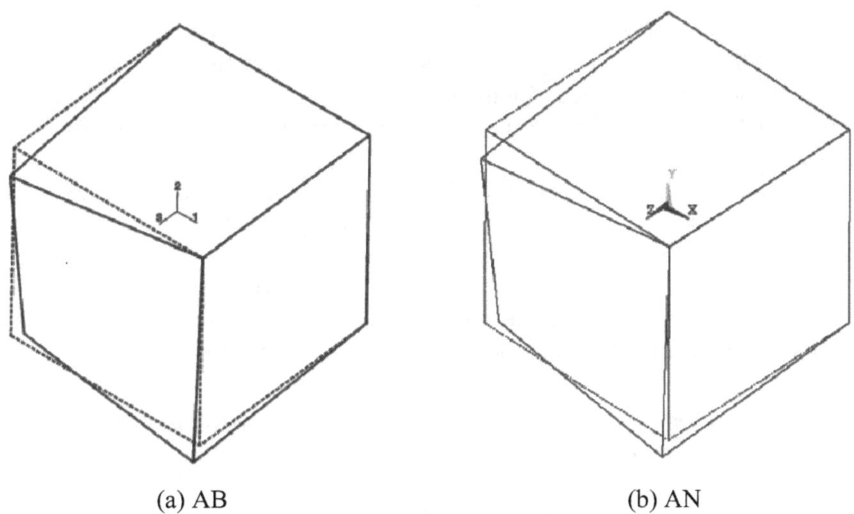

(a) AB (b) AN

Figure 11.2. Residual element distortion after 10 deformation cycles $ABCDEFA$ for the inelastic elements (AB) and (AN). The displacements have not been enhanced.

The results in Table 11.1 indicate that the elements (AD) and (F) exhibit hyperelastic response for the paths considered, while the elements (AB) and (AN) exhibit hypoelastic response. Although the values of E_A^* in Table 11.1 for the elements (AB) and (AN), based on incompatible modes or enhanced strains, seem relatively small, these errors are cumulative when multiple cycles are performed. Figure 11.2 shows the residual element distortion after 10 deformation cycles $ABCDEFA$. It is emphasized that the displacements in Fig. 11.2 have not been enhanced. Also, it is noted that multiple deformation cycles need to be calculated for problems like rolling tires or vibrating MEMS devices so that these accumulated errors may be quite significant in certain calculations.

12 Example problems of thin structures with irregular element shapes

The deformation field associated with the solution of a practical problem typically is inhomogeneous so that the response of the CPE is influenced by the specific form of the inhomogeneous strain energy being used. Mesh refinement tends to cause the response of the CPE to be dominated by its response to homogeneous deformations with the influence of the inhomogeneous strain energy becoming negligible. Consequently, since the CPE satisfies the patch test the predictions of the CPE should converge to the exact solution with mesh refinement. However, the rate of convergence is influenced by details of the functional form for the inhomogeneous strain energy.

In order to study the accuracy of the inhomogeneous strain energy function it is best to focus attention on problems that are dominated by inhomogeneous deformations. This can be accomplished by focusing on the response of thin structures to bending fields. More details of the examples discussed in this section can be found in (Jabareen and Rubin (2007a), Jabareen and Rubin (2007b) and Jabareen and Rubin (2008a)).

12.1 *Shear load on a thin cantilever beam small deformations)*

Figure 12.1 shows a sketch of a thin cantilever beam with dimensions

$$L = 200\,mm \quad , \quad H = W = 10\,mm \tag{12.1}$$

which is fully clamped at one of its ends and is subjected to a shear force P (modeled by a uniform shear stress) applied in the $\mathbf{e}_2$ direction to its other end. The lateral surfaces are traction free. The mesh $\{20n \times n \times n\}$ is defined by distorting the middle cross-section in its reference configuration (using the parameters a_1, a_2, a_3, a_4 shown in Figure 12.1), with $10n$ elements on each side of this cross-section and n elements in each of the $\mathbf{e}_2$ and $\mathbf{e}_3$ directions.

Two cases of element irregularity are considered

$$\begin{aligned} &\text{Case I}: a_1 = a \quad , \quad a_2 = \quad a \quad , \quad a_3 = a \quad , \quad a_4 = \quad a \ , \\ &\text{Case II}: a_1 = a \quad , \quad a_2 = a \quad , \quad a_3 = \quad a \quad , \quad a_4 = \quad a \end{aligned} \tag{12.2}$$

where the parameter a/H defines the element irregularity. Both of these cases cause the middle surface to remain planar with the normal to that surface being in the $\mathbf{e}_1$ $\mathbf{e}_2$ plane for Case I and being in the $\mathbf{e}_1$ $\mathbf{e}_3$ plane for Case II. The value

$$u^*_{A2} = 0.21310\,mm \quad for \quad P = 0.1\,N \tag{12.3}$$

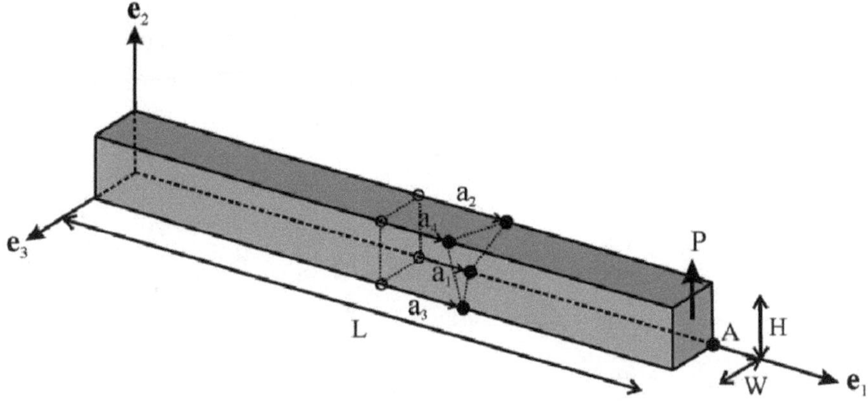

Figure 12.1. Shear load on a thin cantilever beam. The irregular element mesh is based on the distorted center cross-section.

of the $\mathbf{e}_2$ component of the displacement of point A (see Figure 12.1) predicted by the generalized CPE (C) with the most refined mesh ($n = 5$) and zero irregularity ($a/H = 0$) is considered to be exact and the error E associated with the predictions u_{A2} of other calculations for the same value of P is defined by

$$E = \frac{u_{A2} \quad u_{A2}^*}{|u_{A2}^*|} \tag{12.4}$$

Figures 12.2a,b show the results for Case I and Figs. 12.2c,d show the results for Case II. The error is plotted as a function of the irregularity parameter a/H in Figs. 12.2a,c and convergence is examined in Figs. 12.2b,d. Ideally the response should be insensitive to the value of a/H. These figures show that the two elements converge to the same value for the refined mesh ($n = 5$) and large irregularity $a/H = 2$. They also show that the predictions of (C) are slightly more accurate than those of (F).

12.2 *Shear load on a thin slanted cantilever beam small deformations)*

Figure 12.3 shows a sketch of a thin slanted cantilever beam with dimensions (12.1) and with the slanting angle θ. The boundary conditions are the same as those for the previous example except that the shear load P is applied in the $\mathbf{e}_3$ direction to cause out-of-plane bending. Again the mesh is taken to be $\{20n \times n \times n\}$ with $20n$ elements in axial direction of the

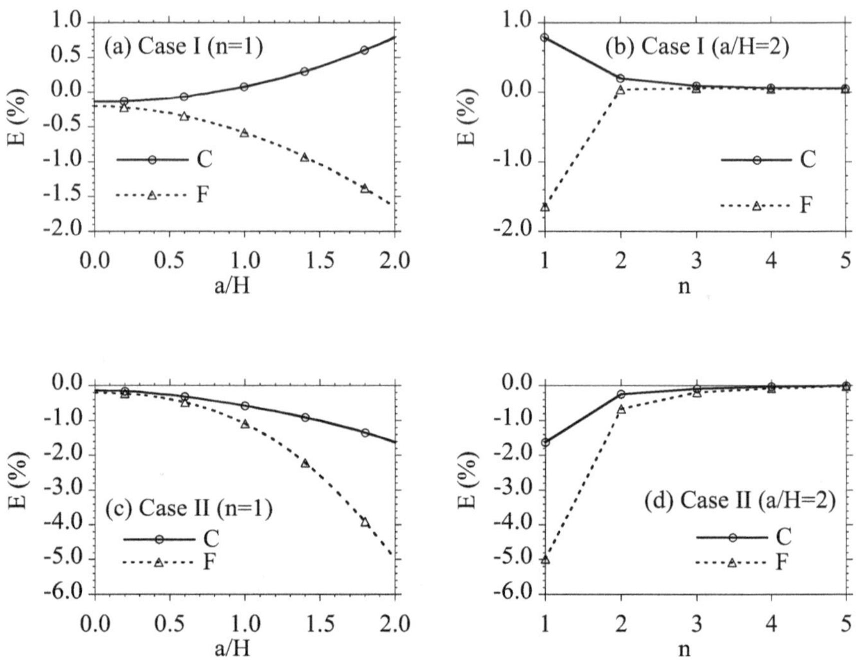

Figure 12.2. Shear load on a thin cantilever beam (small deformations). (a,c) Errors in the displacement of the point A in the $\mathbf{e}_2$ direction versus the distortion parameter a/H and; (b,d) the errors versus n for the mesh $\{20n \times n \times n\}$ defined for two cases of element distortion.

beam. All of the elements have parallelogram cross-sections in the $\mathbf{e}_1$ $\mathbf{e}_2$ plane with sides parallel to the ends of the beam.

Figure 12.4a shows the displacement component u_{A3} of point A (see Figure 12.3) in the $\mathbf{e}_3$ direction as a function of θ for the most refined mesh ($n = 5$). The error E in u_{A3} is defined in a similar manner to (12.4) with the exact value u^*_{A3} taken to be that predicted by (C) for each value of θ with $n = 5$ and with the load P given by (12.3). Figures 12.4b,c show that (C) and (F) converge to the same values and that (C) is slightly more accurate than (F) for $n = 1$ and large values of θ.

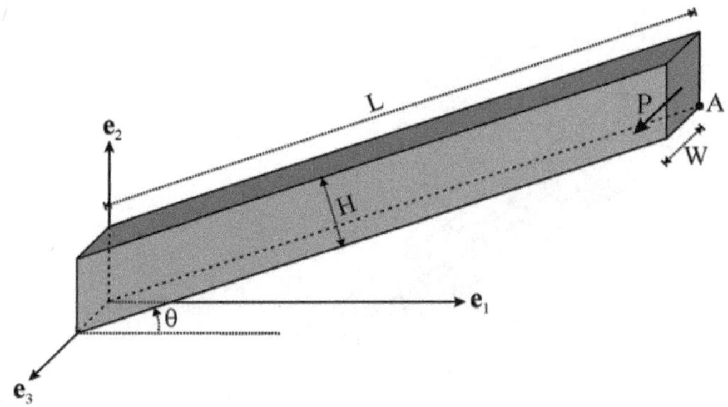

Figure 12.3. Shear load on a thin slanted cantilever beam.

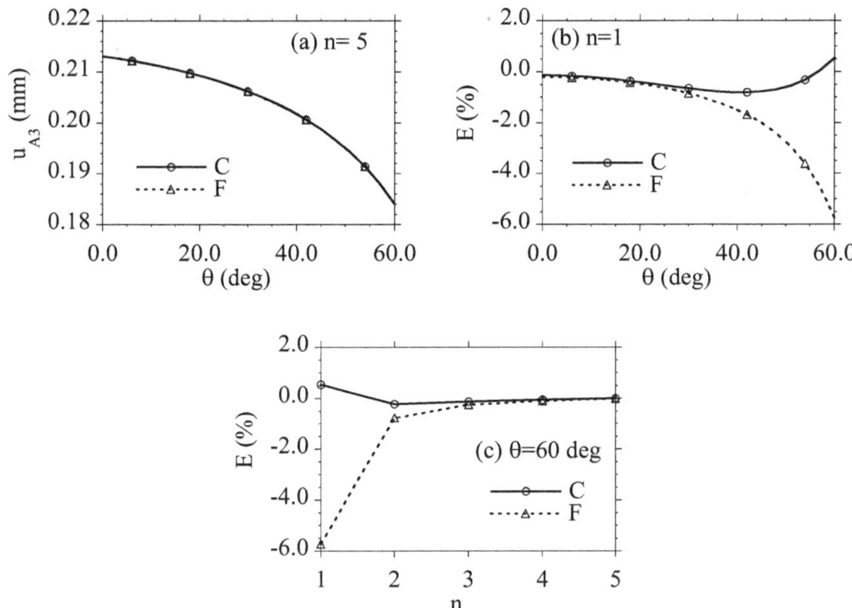

Figure 12.4. Shear load on a thin slanted cantilever beam (small deformations). (a) Displacement u_{A3} of the point A in the $\mathbf{e}_3$ direction versus the angle θ for $n = 5$ with the mesh $\{20n \times n \times n\}$; (b) errors in u_{A3} versus θ for $n = 1$; (c) errors in u_{A3} versus n for $\theta = 60^{\circ}$.

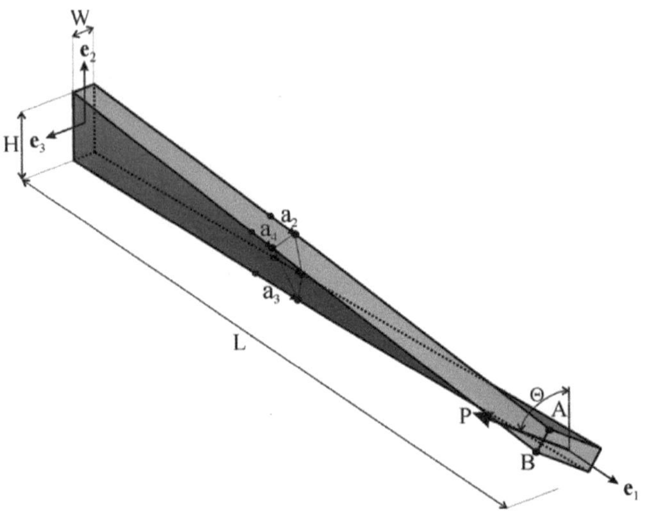

Figure 12.5. Shear load on a thin twisted cantilever beam. The element mesh is based on the distorted center cross-section.

12.3 *Lateral torsional buckling of a thin cantilever beam large deformations)*

Figure 12.5 shows a sketch of a thin twisted cantilever beam which in its unstressed reference configuration has length L and a rectangular cross-section with height H and width W given by

$$L = 200\,mm \quad , \quad H = 10\,mm \quad , \quad W = 2\,mm \tag{12.5}$$

Each of the cross-sections is twisted by the angle θ which varies linearly from zero at the clamped end to Θ at the loaded end. Also, the load P is applied in the constant direction parallel to the long edges of the rectangular cross-section in its reference configuration. In order to stimulate lateral torsional buckling the value of Θ is taken to be 0.1° which introduces a small imperfection in the reference geometry of the beam. Furthermore, the element irregularity shown in Figure 12.5 is specified by Case I in (12.2) and the mesh is given by $\{20n \times n \times n\}$ with $20n$ elements in axial direction of the beam.

To investigate rotation of the beam's end it is convenient to consider the difference in the displacements of the points A and B shown in Figure 12.5. Specifically, the quantity u is defined by

$$u = (\mathbf{u}_B \quad \mathbf{u}_A) \bullet \mathbf{e}_2 \tag{12.6}$$

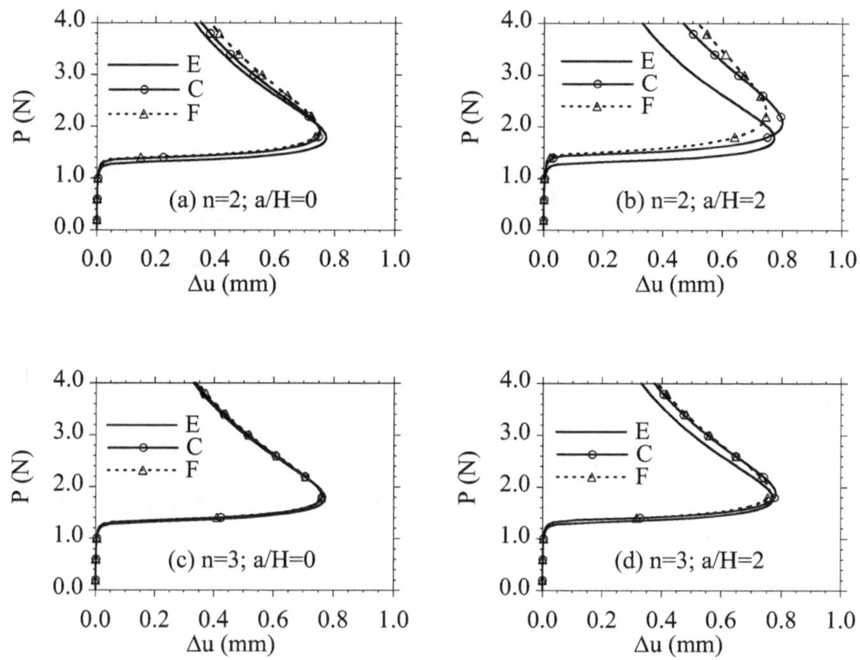

Figure 12.6. Large deformation lateral torsional buckling of a thin cantilever beam with a small pre-twist $\Theta = 0.1^{\circ}$ using the mesh $\{20n \times n \times n\}$. The influence of element irregularity is shown in Figs. 12.6a,b for $n = 2$ and in Figs. 12.6c,d for $n = 3$.

Figure 12.6 shows the results for large deformation lateral torsional buckling of a thin cantilever beam. In this figure the curves denoted by (E) are predicted by (C) with $n = 5$ and $a/H = 0$ and are considered to be exact. The results in this figure show that for $n = 2$ the predictions are not yet converged and are sensitive to element irregularity whereas for $n = 3$ the predictions are reasonably converged and reasonably insensitive to element irregularity. Also, it can be seen that (C) and (F) converge to the same results.

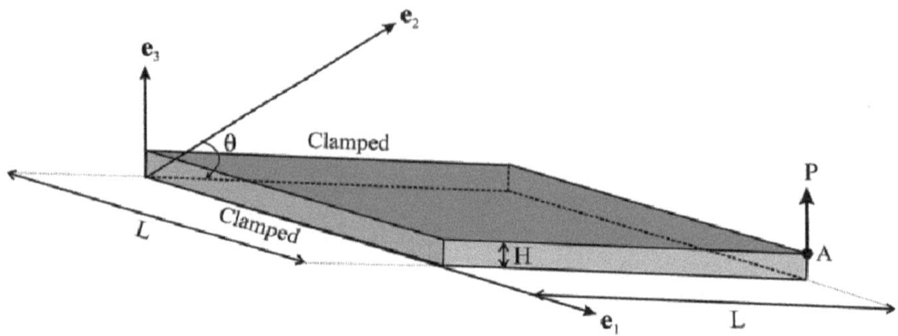

Figure 12.7. Point load on the corner of a thin partially clamped rhombic plate.

12.4 *Point load on the corner of a thin partially clamped rhombic plate small deformations)*

Figure 12.7 shows a sketch of a thin partially clamped rhombic plate with dimensions

$$L = 500\,mm \quad , \quad H = 10\,mm \tag{12.7}$$

with two clamped and two free edges and which is loaded at its corner by a point force P. The length of each edge is L and the load is specified by

$$P = 1\,N \tag{12.8}$$

The mesh used for the plate is defined by $\{10n \times 10n \times n\}$ with n elements through the thickness.

Figure 12.8a shows the component u_{A3} of the displacement of the point A in the $\mathbf{e}_3$ direction as a function of θ for the most refined mesh ($n = 5$). The error E in this displacement is defined in a similar manner to (12.4) with the exact value u^*_{A3} taken to be that predicted by (C) for each value of θ with $n = 5$ and the load P given by (12.8). Also, Figures 12.8b,c show that (C) and (F) converge to the same values.

12.5 *Point load on the center of a thin fully clamped square plate with an irregular element mesh small deformations)*

Figure 12.9 shows a sketch of one quarter of a thin fully clamped square plate with dimensions (12.7) that is loaded by a point force at its center. Only one quarter of the plate is modeled and the value P given by (12.8) corresponds to one quarter of the load that would be applied to the center

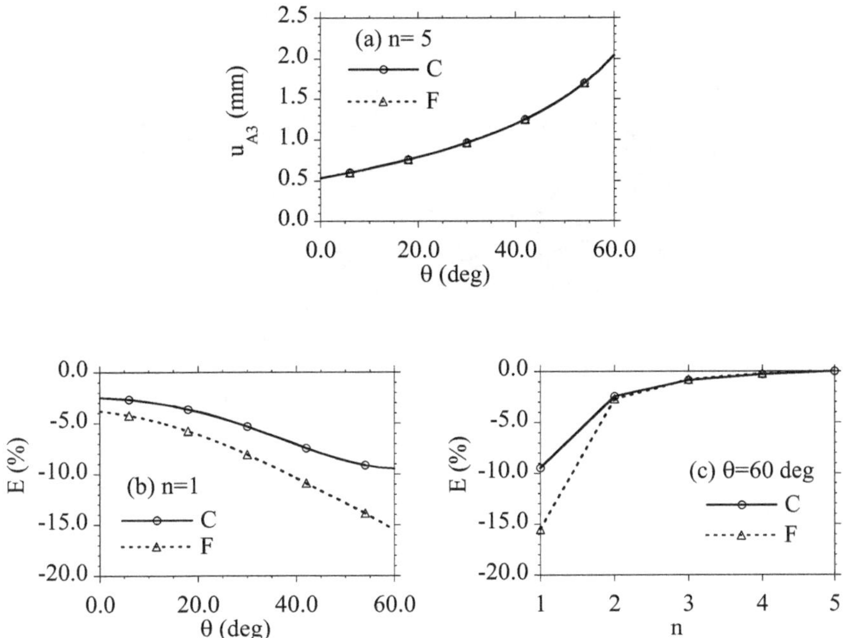

Figure 12.8. Point load on the corner of a thin partially clamped rhombic plate (small deformations). Displacement u_{A3} of the point A in the $\mathbf{e}_3$ direction versus the angle θ for $n = 5$ with the mesh $\{10n \times 10n \times n\}$; (b) errors in u_{A3} versus θ for n=1; (c) errors in u_{A3} versus n for $\theta = 45^o$.

of the entire plate. Irregular elements are specified by moving the center point of the quarter section to the position characterized by the lengths a_1 and a_2 (shown in Figure 12.9) defined by two cases

$$\text{Case I} : a_1 = a_2 = a \ , \quad 1 \leq \frac{4a}{L} \leq 1 \ ,$$
$$\text{Case II} : a_1 = \frac{L}{4} cos(\theta) \ , \quad a_2 = \frac{L}{4} sin(\theta) \ , \quad 0 \leq \theta \leq 2\pi \tag{12.9}$$

The quarter section of the plate is meshed by $\{10n \times 10n \times n\}$ with each subsection being meshed by $\{5n \times 5n \times n\}$ and with n elements through the thickness. The error E in the displacement component u_{A3} of point A in the $\mathbf{e}_3$ direction is defined in a similar manner to (12.4) with the exact value u^*_{A3} taken to be that predicted by (C) for regular elements $(a/L = 0)$

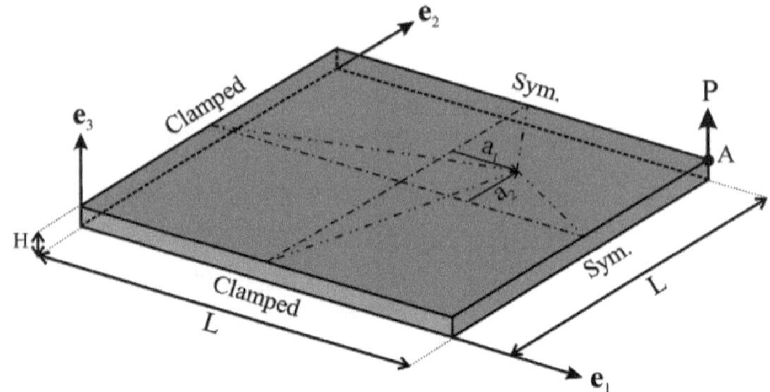

Figure 12.9. Point load on the center of a thin fully clamped square plate with an irregular element mesh.

with $n = 5$

$$u^*_{A3} = 0.16893\,mm \quad for \quad P = 1\,N \tag{12.10}$$

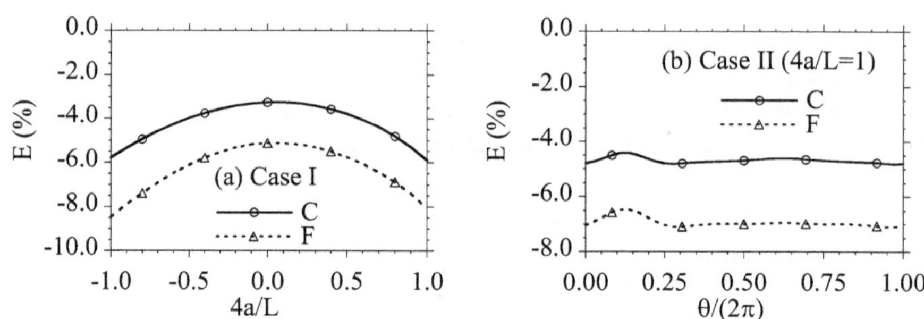

Figure 12.10. Point load on the center of a thin fully clamped square plate (small deformations). Errors in the displacement of the point A in the $\mathbf{e}_3$ direction versus the distortion parameters (a) $4a/L$ and; (b) the angle θ for two cases of element irregularity with the mesh $\{10 \times 10 \times 1\}$.

Figures 12.10a,b show the error for $n = 1$ as a function of the irregularity parameter $4a/L$ for Case I (Figure 12.10a) and as a function of $\theta/(2\pi)$ for Case II (Figure 12.10b). From these figures it can be seen that that (C) and (F) are both relatively insensitive to the magnitude and type of element irregularity.

12.6 *Point load on the corner of a thin partially clamped rhombic plate large deformations)*

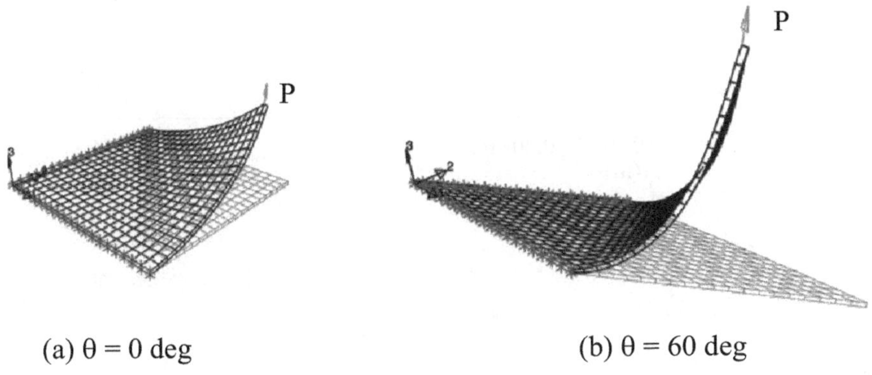

Figure 12.11. Point load on a partially clamped rhombic plate (large deformations). Predictions of the generalized CPE for the mesh $\{10n \times 10n \times 1\}$ with $n = 2$ and $P = 1\,kN$.

Figure 12.11 shows the deformed shapes of a thin partially clamped rhombic plate subjected to a point load on its corner for two different angles θ and the same value of the load P. The plate is fully clamped on two edges and the other edges and major surfaces are traction free. The dimensions are given by (12.7) as shown in Figure 12.7 and the point force P given by

$$P = 1\,kN \tag{12.11}$$

The mesh is specified by $\{10n \times 10n \times 1\}$ and the exact values u_3^* of the displacement of the corner in the $\mathbf{e}_3$ direction is determined by the most refined solution (C) with $n = 5$

$$\begin{aligned} u_3^* &= 0.21084\,m \quad for \quad \theta = 0^\circ \ , \\ u_3^* &= 0.39306\,m \quad for \quad \theta = 60^\circ \end{aligned} \tag{12.12}$$

Figures 12.12 show the load P versus displacement curves for $n = 2$ and the convergence curves for two values of the angle θ. Comparison of Figures 12.12a,c shows that the rhombic plate with angle $\theta = 60^\circ$ is more flexible than that for $\theta = 0^\circ$ and that (C) and (F) converge to the same solution. Also, Figure 12.12d shows that the convergence properties of (C) are slightly better than those of (F) for the case when $\theta = 60^\circ$.

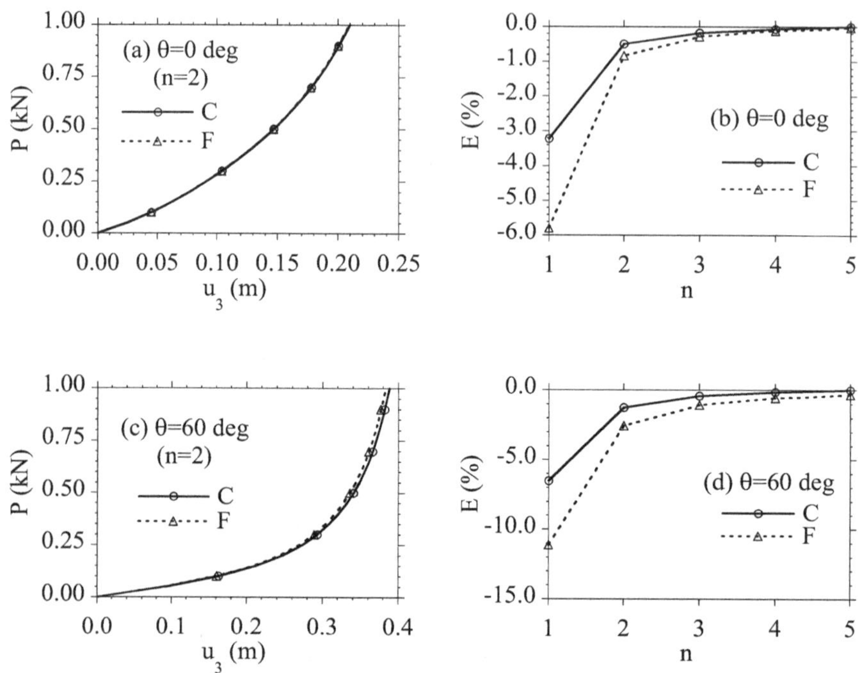

Figure 12.12. Point load on a partially clamped rhombic plate (large deformations). Predictions of the load P versus displacement u_3 at the loaded corner and convergence of the error in the displacement for the mesh $\{10n \times 10n \times 1\}$ with the load $P = 1\ kN$ and different angles.

13 Example problems exhibiting robustness to hourglass instabilities

Reese and Wriggers (1996), Reese and Wriggers (2000) and Reese et al. (2000) have shown that enhanced strain formulations, like that proposed by Simo et al. (1993), can predict unphysical hourglass buckling modes for plane strain compression of a block. In order to examine this phenomena (Jabareen and Rubin (2007a)) considered a square block with edge length $L = 1\ m$ which is compressed between two smooth rigid parallel end plates with the other two edges being free (see Figure 13.1). Plane strain deformations are modeled using one 3-D element through the block's thickness and eliminating displacements in the out-of-plane direction.

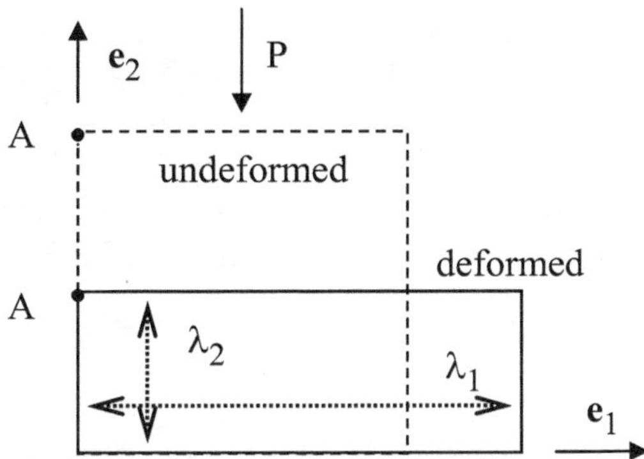

Figure 13.1. Plane strain compression of a block showing the load P and the stretches λ_1 and λ_2 for the homogeneous solution.

For the homogeneous solution the stretches in the $\mathbf{e}_1$ and $\mathbf{e}_2$ directions are denoted by λ_1 and λ_2, respectively (see Figure 13.1). At a critical value of λ_2 (< 1) the block buckles in its plane. In order to calculate the post-buckling response of this structure it is necessary to use special methods like arc-length control because a spring-back phenomena occurs as the block buckles in shear. Moreover, a small imperfection is introduced in the reference mesh to trigger the shear buckling mode. Figure 13.2 shows that (C) predicts physical shear buckling modes for two nearly perfect regular meshes $\{10 \times 10 \times 1\}$ and $\{20 \times 20 \times 1\}$ and for two nearly perfect irregular meshes. From this figure it can be seen that the effect of element irregularity is not large.

Figure 13.3 shows the predictions of an element in ABAQUS (AB) which is based on reduced integration with hourglass control. From this figure it can be seen that this element (AB) produces physical shear buckling which follows the predictions of the Cosserat element (C). However, (AB) ceases to converge and thus cannot predict the full post-buckling behavior. Moreover, it is noted that the buckled mode predicted by (AB) shown in Figure 13.3 is presented for the load just before the program ceased to converge.

As mentioned previously, the elements based on enhanced strains and incompatible modes can exhibit unphysical hourglass modes for problems with high compression combined with bending. Specifically, Figure 13.4 shows the results of calculations using the enhanced strain element in FEAP (F) for two perfect regular meshes $\{10 \times 10 \times 1\}$ and $\{20 \times 20 \times 1\}$. From this figure it can be seen that at the bifurcation point $u_{A2} =\ \ 0.330\,m$ (the

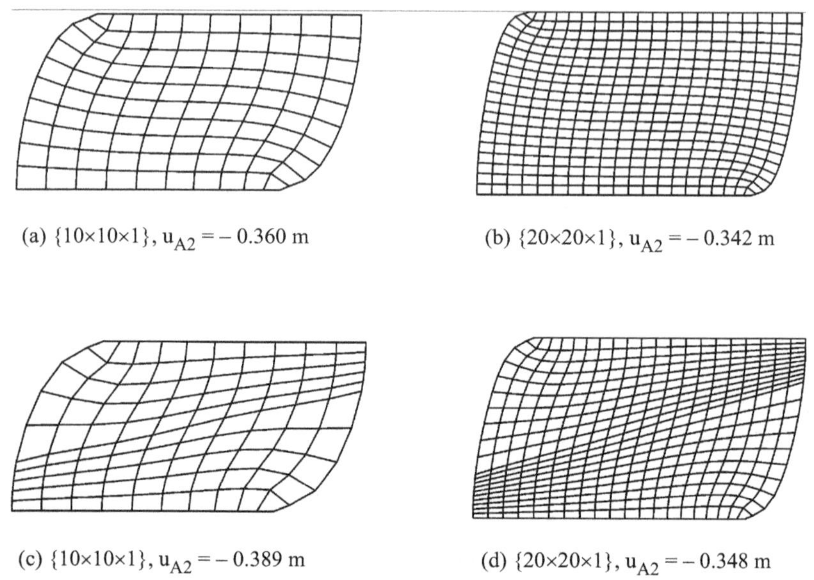

Figure 13.2. Compression of a block. Physical shear buckling modes predicted by (C) for: (a,b) two nearly perfect regular meshes $\{10 \times 10 \times 1\}$ and $\{20 \times 20 \times 1\}$; and (c,d) two nearly perfect irregular meshes.

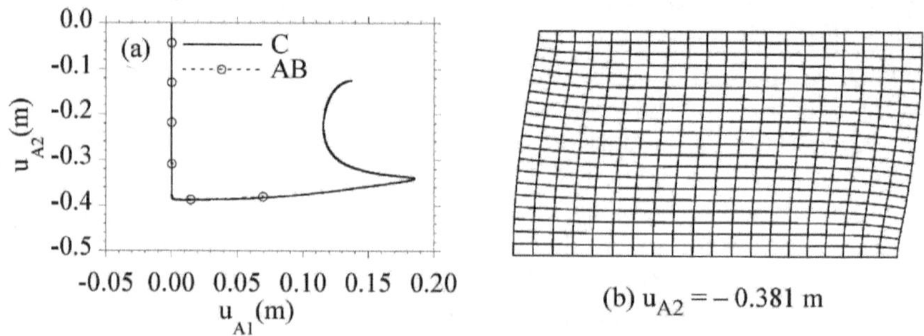

Figure 13.3. Compression of a block. Predictions of (C) and (AB) in ABAQUS for a nearly perfect regular mesh $\{20 \times 20 \times 1\}$; (a) displacement components; (b) compressive force P; and (b) post-buckling shape.

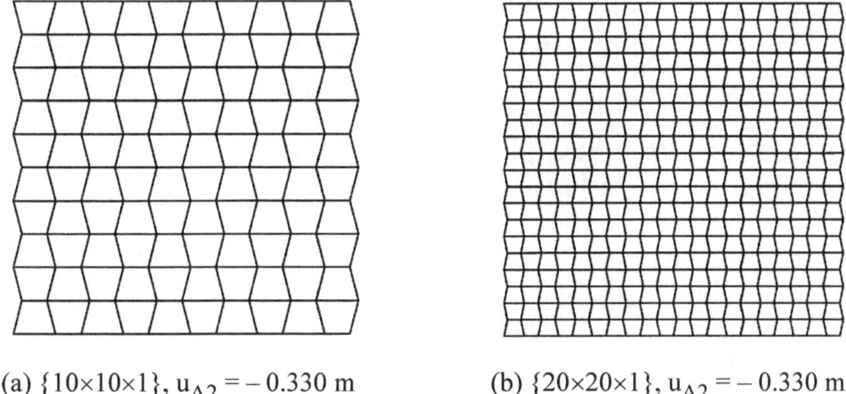

(a) $\{10\times10\times1\}$, $u_{A2} = -0.330$ m (b) $\{20\times20\times1\}$, $u_{A2} = -0.330$ m

Figure 13.4. Compression of a block. Unphysical hourglass buckling modes at bifurcation predicted by (F) for two perfect regular meshes $\{10 \times 10 \times 1\}$ and $\{20 \times 20 \times 1\}$.

point where the lowest eigenvalue of the global tangent stiffness changes sign between $u_{A2} = \quad 0.3303\, m$ and $u_{A2} = \quad 0.3304\, m$), the associated buckling mode shapes are characterized by unphysical hourglassing for both meshes.

The enhanced strain and incompatible mode elements in ABAQUS (AB), ADINA (AD) and ANSYS (AN) exhibit unphysical hourglasing that causes lack of convergence for (AB) and (AN). Figure 13.5 shows the deformed shapes predicted by these elements for a nearly perfect regular mesh $\{20 \times 20 \times 1\}$ corresponding to the load just before the programs ABAQUS and ANSYS ceased to converge. The element (AD) predicts a post buckled response that is corrupted by hourglassing.

14 Example problems exhibiting robustness to near incompressibility

Jabareen and Rubin (2008b) considered the example of plane strain indentation of a rigid plate into a nearly incompressible elastic block to examine the response of (C) in the nearly incompressible limit. Figure 14.1 shows a sketch of the boundary conditions on a block which has length $2L$, height L and depth W. Material points on the block's sides and bottom remain in contact with a rigid container and are allowed to slide freely. The top surface of the block is loaded by a rigid plate (AB) of length L which makes

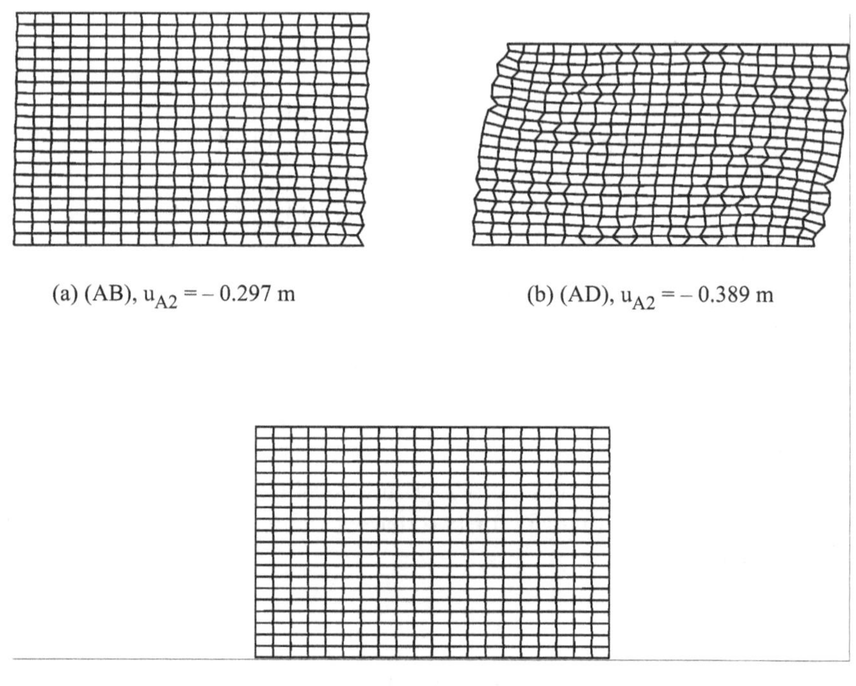

Figure 13.5. Compression of a block. Unphysical hourglass buckling modes predicted by (AB), (AD) and (AN) for a refined nearly perfect regular mesh $\{20 \times 20 \times 1\}$.

perfect contact with the block so that material points in contact with the rigid plate move only vertically. The remaining half of the block's top surface is traction free and the dimensions of the block are given by

$$L = W = 1\,m \tag{14.1}$$

Irregular meshes are defined by dividing the block into four subsections with the central node moving to the position characterized by the lengths

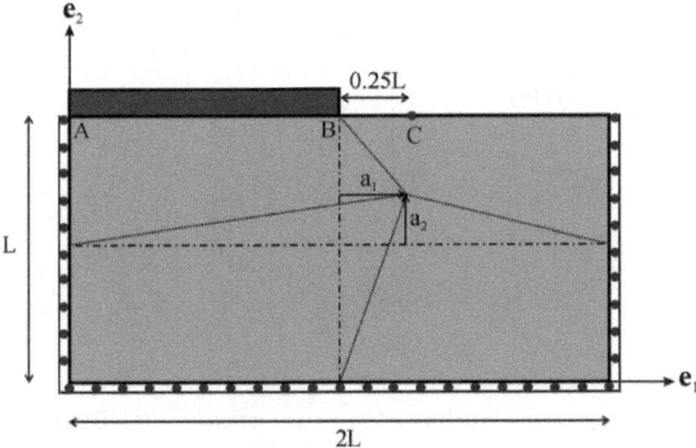

Figure 14.1. Plane strain indentation of a rigid plate into a block showing the boundary conditions and definition of element irregularity.

$\{a_1, a_2\}$ (shown in Figure 14.1) defined by two cases

$$\begin{aligned} \text{Case I}: a_1 &= a \ , \quad a_2 = 0 \ , \quad 1 \leq \frac{8a}{3L} \leq 1 \ , \\ u_{A2} &= \quad 0.1\, m \ , \quad n = 5 \ , \\ \text{Case II}: a_1 &= 0 \ , \quad a_2 = a \ , \quad 1 \leq \frac{8a}{3L} \leq 1 \ , \\ u_{A2} &= \quad 0.1\, m \ , \quad n = 5 \end{aligned} \tag{14.2}$$

The entire block is meshed by $\{8n \times 4n \times 1\}$ with $4n$ elements in the $\mathbf{e}_1$ direction and $2n$ elements in the $\mathbf{e}_2$ in each of the subsections. The point C (shown in Figure 14.1) is located on the free top surface at a distance $0.25L$ from the corner B of the rigid plate. Also, the material is considered to be nearly incompressible.

Figure 14.2 shows convergence of the solution for the regular $(a = 0)$ mesh $\{8n \times 4n \times 1\}$ and $u_{A2} = \quad 0.1\, m$. The converged value u^*_{C2} of the displacement of the point C in the $\mathbf{e}_2$ direction predicted by (C) for a regular mesh with $n = 20$ is considered to be exact and is given by

$$u^*_{C2} = 0.071895\, m \quad for \quad u_{A2} = \quad 0.1\, m \quad with \quad n = 20 \tag{14.3}$$

The error E of in the values u_{C2} predicted by calculations of other elements and meshes is defined by an expression similar to (12.4). Figure 14.2 shows the convergence of this error predicted by (C), (Q1P0) and (HO9), where (HO9) denotes a mixed higher order nine node quadrilateral element in

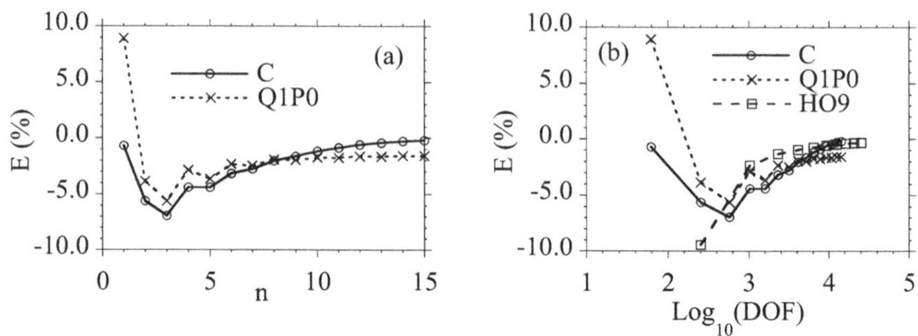

Figure 14.2. Plane strain indentation of a rigid plate into a nearly incompressible block. Convergence of the error E in the displacement u_{C2} of the point C using the regular mesh $\{8n \times 4n \times 1\}$ for $u_{A2} = \quad 0.1\,m$ versus: (a) n; and (b) versus the number of degrees of freedom DOF.

FEAP. This error is plotted relative to n for the mesh $\{8n \times 4n \times 1\}$ in Figure 14.2a and is plotted relative to the degrees of freedom (DOF, calculated for plane strain response) in Figure 14.2b. From Figure 14.2a it is not clear if (Q1P0) exhibits a locking behavior by converging to a value different from (C) or whether the convergence rate is very slow. To validate the converged value of (C) for $n = 20$, calculations were also performed using the mixed higher order element (HO9) with the mesh $\{8n \times 4n \times 1\}$ up to $n = 10$. In particular, it can be seen in Fig. 14.2b that (HO9) tends to converge to the value predicted by (C).

Figure 14.3 presents the errors E in the displacement u_{C2} for two cases of element irregularity and for the mesh $\{8n \times 4n \times 1\}$ with $n = 5$ and $u_{A2} = \quad 0.1\,m$. Since there is a strain concentration near the edge of the plate it is expected that a non-fully converged solution will be sensitive to element irregularity. In particular, it can be seen from Figure 14.3a that (Q1P0) is more sensitive to element irregularity than (C) for positive values of a for Case I which cause the elements near the plate's edge B to be more irregular. The results in Figure 14.3b show that the error reduces slightly for increasing positive values of a for Case II which cause the elements near the plate's edge B to be more refined.

Figure 14.4 shows nonlinear load curves using the regular mesh $\{8n \times 4n \times 1\}$ for different values of n. These figures again show that (C) predicts more flexible response than (Q1P0) for the coarser meshes. Figure 14.5 shows the deformed shapes for the regular mesh $\{8n \times 4n \times 1\}$ with $n = 3$ for different values of loads. In particular, it can be seen that the flexibility

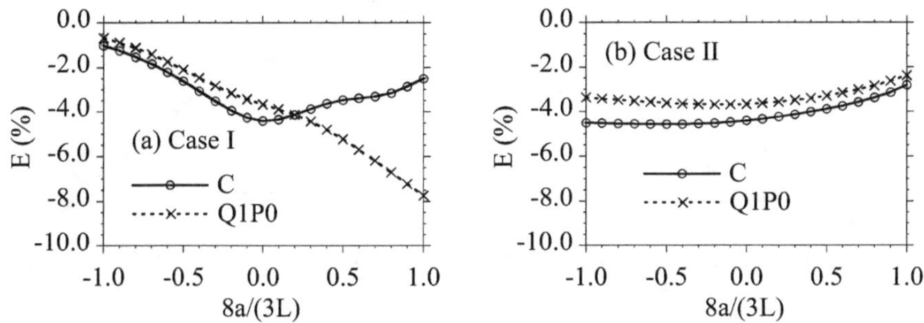

Figure 14.3. Plane strain indentation of a rigid plate into a nearly incompressible block. Error E in the displacement u_{C2} of the point C for two cases of element irregularity and for the mesh $\{8n \times 4n \times 1\}$ with $n = 5$ and $u_{A2} = \quad 0.1\, m$.

of (C) allows the elements near the plate's corner to roll around the corner more easily than allowed by (Q1P0). Since the flexibility of (C) has been validated relative to the mixed higher order element (HO9) it is concluded that the stiffness shown by (Q1P0) is unphysical.

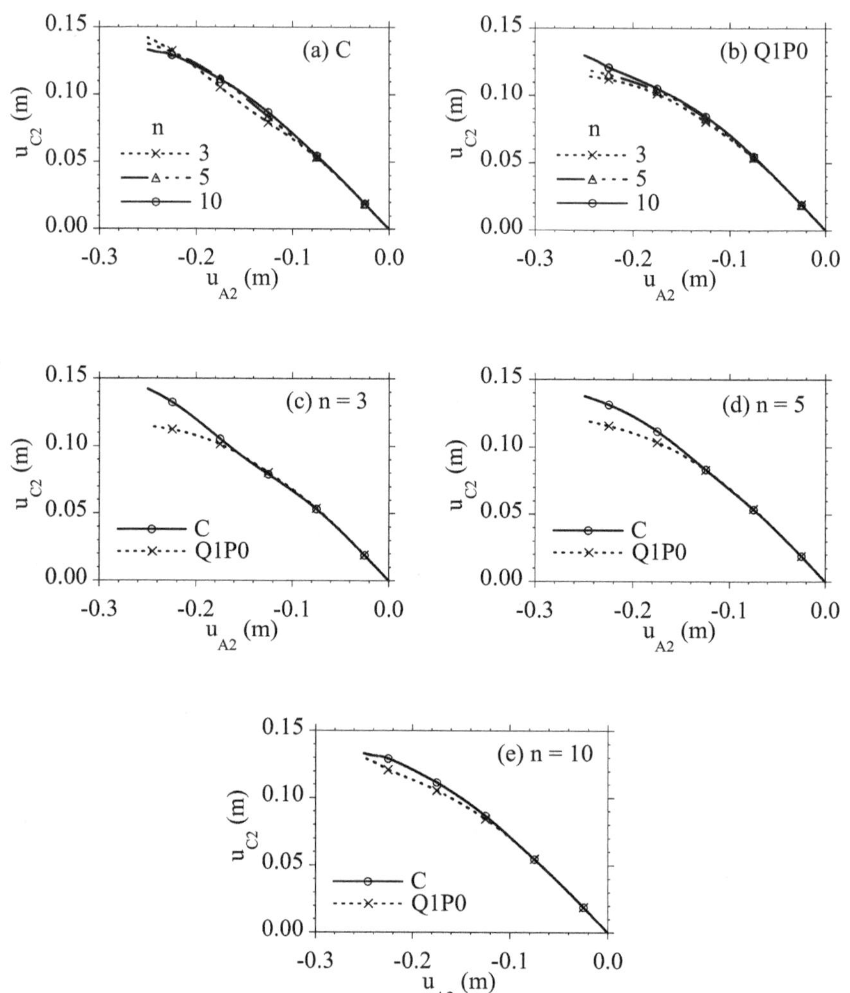

Figure 14.4. Plane strain indentation of a rigid plate into a nearly incompressible block showing nonlinear load curves using the regular mesh $\{8n \times 4n \times 1\}$ for different values of n.

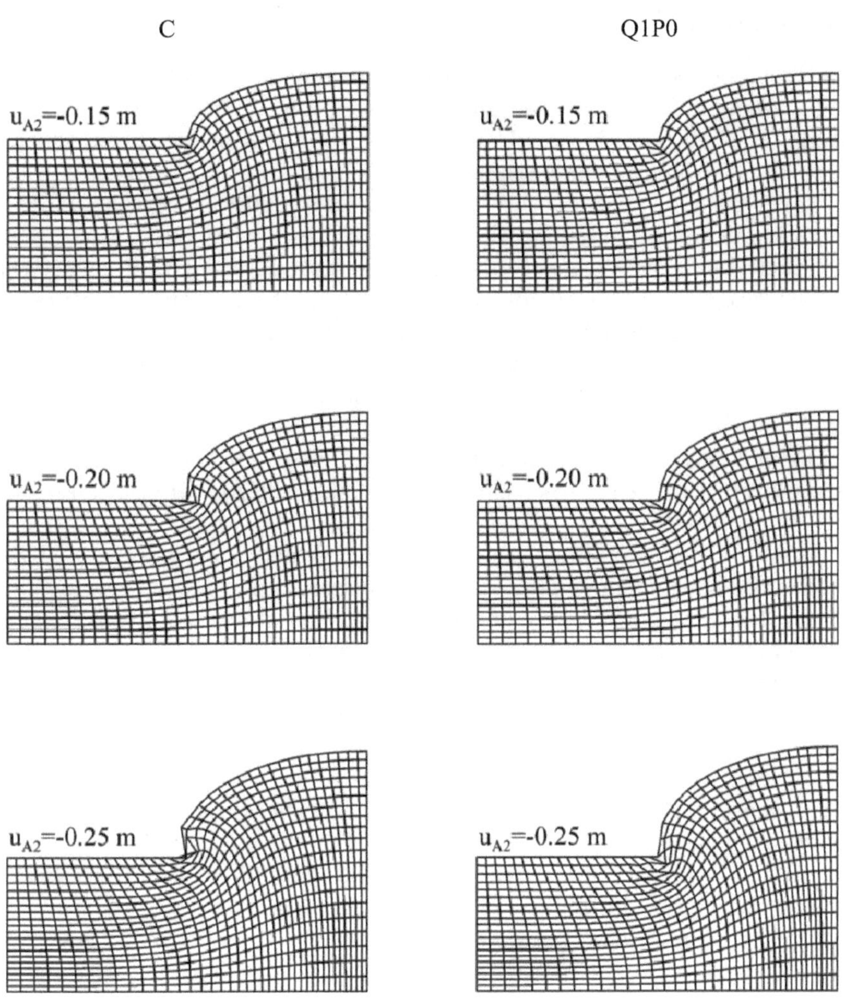

Figure 14.5. Plane strain indentation of a rigid plate into a nearly incompressible block showing the deformed shapes for the regular mesh $\{8n \times 4n \times 1\}$ with $n = 5$. The left column shows the results for (C) and the right column shows the results (Q1P0).

15 Conclusions

The previous sections summarized the development of a 3-D brick Cosserat Point Element (CPE) for the numerical solution of problems in nonlinear elasticity. The CPE is based on the theory of a Cosserat point which is a continuum theory that introduces balance laws for the deformation of a structure that is "thin" in three-dimensions. In contrast with standard finite element methods, the CPE treats the finite element as a structure and the kinetic quantities are determined by derivatives of a strain energy that characterizes resistance to all modes of deformation of the structure. In particular, a nonlinear form of the patch test is used to place restrictions on this strain energy function which ensure that the CPE reproduces all homogeneous solutions exactly for all reference element shapes. Special attention has been focused on developing an analytical form for the strain energy of inhomogeneous deformations that causes the predictions of the CPE to be relatively insensitive to element irregularity even for thin structures like shells and rods.

Example problems have been considered which show that the CPE is as accurate as elements based on enhanced strains and incompatible modes for thin structures and is free of the hourglass instabilities observed in these elements for deformations with high compression combined with bending. Also, the CPE is free of locking due to near incompressible material response. Consequently, the CPE is truly a user friendly element that can be used with confidence to solve problems in nonlinear elasticity.

Although the CPE approach has proved very successful for nonlinear elastic materials it is not clear how it can be generalized for more complicated material response. An elastic solid has the special simple property that it has a unique shape when it is unloaded. In contrast, an elastic-viscoplastic material can have an infinite number of stress-free shapes which differ by a general homogeneous deformation. Thus, in order to generalize the CPE for elastic-viscoplastic materials it is necessary to first understand how the CPE can be generalized for fluids which have no unique stress-free shapes. At present it appears that this area of research will remain challenging for a number of years to come.

Bibliography

ABAQUS. Inc., Version 6.5-1, Providence RI 02909-2499.

ADINA. Inc., Version 8.3.1, Watertown M 02472.

ANSYS. Inc., University dvanced Version 9 Canonsburg, P 15317.

T. Belytschko, J.S.J. Ong, W.K. Liu, and J.M. Kennedy. Hourglass control in linear and nonlinear problems. *Comp. Meth. Appl. Mech. Engrg.*, 43: 251–276, 1984.

E.F.I. Boerner, S. Loehnert, and P. Wriggers. A new finite element based on the theory of a Cosserat point - extension to initially distorted elements for 2D plane strain. *Int. J. Numer. Meth. Engng.*, 71:454–472, 2007.

FEAP. - Finite Element nalysis Program, Version 7.5, University of California, Berkeley.

P. Flory. Thermodynamic relations for high elastic materials. *Trans. Faraday Soc.*, 57:829–838, 1961.

R. Hutter, P. Hora, and P. Niederer. Total hourglass control for hyperelastic materials. *Comp. Meth. Appl. Mech. Engrg.*, 189:991–1010, 2000.

M. Jabareen and M.B. Rubin. Hyperelasticity and physical shear buckling of a block predicted by the Cosserat point element compared with inelasticity and hourglassing predicted by other element formulations. *Computational Mechanics.*, 40:447–459, 2007a.

M. Jabareen and M.B. Rubin. An improved 3-D Cosserat brick element for irregular shaped elements. *Computational Mechanics.*, 40:979–1004, 2007b.

M. Jabareen and M.B. Rubin. Modified torsion coefficients for a 3-D brick Cosserat point element. *Computational Mechanics.*, 41:517–525, 2007c.

M. Jabareen and M.B. Rubin. A generalized Cosserat point element (CPE) for isotropic nonlinear elastic materials including irregular 3-D brick and thin structures. *Journal of Mechanics of Materials and Structure*, 3 (8): 1465–1498, 2008a.

M. Jabareen and M.B. Rubin. A Cosserat point element (CPE) for nearly planar problems (including thickness changes) in nonlinear elasticity. *International Journal for Engineering Science.*, 46 (10):989–1010, 2008b.

S. Loehnert, E.F.I. Boerner, M.B. Rubin, and P. Wriggers. Response of a nonlinear elastic general Cosserat brick element in simulations typically exhibiting locking and hourglassing. *Computational Mechanics.*, 36:255–265, 2005.

B. Nadler and M.B. Rubin. A new 3-D finite element for nonlinear elasticity using the theory of a Cosserat point. *Int. J. Solids and Structures.*, 40: 4585–4614, 2003.

P.M. Naghdi and M.B. Rubin. Restrictions on nonlinear constitutive equations for elastic shells. *J. Elasticity.*, 39:133–163, 1995.

S. Reese and P. Wriggers. Finite element calculation of the stability behaviour of hyperelastic solids with the enhanced strain methods. *Zeitschrift fur angewandte Mathematik und Mechanik.*, 76:415–416, 1996.

S. Reese and P. Wriggers. A stabilization technique to avoid hourglassing in finite elasticity. *Int. J. Numer. Meth. Engng.*, 48:79–109, 2000.

S. Reese, P. Wriggers, and B.D. Reddy. A new locking free brick element technique for large deformation problems in elasticity. *Computers and Structures.*, 75:291–304, 2000.

M.B. Rubin. On the theory of a Cosserat point and its application to the numerical solution of continuum problems. *J. Appl. Mech.*, 52:368–372, 1985a.

M.B. Rubin. On the numerical solution of one-dimensional continuum problems using the theory of a Cosserat point. *J. Appl. Mech.*, 52:373–378, 1985b.

M.B. Rubin. Numerical solution of two- and three-dimensional thermomechanical problems using the theory of a Cosserat point. *J. of Math. and Physics ZAMP).*, 46, Special Issue, S308-S334. In Theoretical, Experimental, And Numerical Contributions To The Mechanics Of Fluids And Solids, Edited by J Casey and MJ Crochet, Brikhauser Verlag, Basel (1995), 1995.

M.B. Rubin. Restrictions on nonlinear constitutive equations for elastic rods. *J. Elasticity.*, 44:9–36, 1996.

M.B. Rubin. *Cosserat Theories: Shells, Rods and Points. Solid Mechanics and its Applications, Vol. 79.* Kluwer, 2000.

M.B. Rubin. Numerical solution procedures for nonlinear elastic rods using the theory of a Cosserat point. *Int. J. Solids Structures.*, 38:4395–4437, 2001.

J.C. Simo and F. Armero. Geometrically non-linear enhanced strain mixed methods and the method of incompatible modes. *Int. J. Numer. Meth. Engng.*, 33:1413–1449, 1992.

J.C. Simo and M.S. Rifai. A class of mixed assumed strain methods and the method of incompatible modes. *Int. J. Numer. Meth. Engng.*, 29: 1595–1638, 1990.

J.C. Simo, F. Armero, and R.L. Taylor. Improved versions of assumed enhanced strain tri-linear elements for 3D finite deformation problems. *Comp. Meth. Appl. Mech. Engrg.*, 110:359–386, 1993.

I.S. Sokolnikoff. *Mathematical Theory of Elasticity.* McGraw-Hill, 1956.

Multiscale Approaches: From the Nanomechanics to the Micromechanics

Esteban P. Busso

Centre des Matériaux, MINES ParisTech
CNRS-UMR 7633
B.P. 87, F-91003 Evry, France

1 Overview

Computational modelling of materials behaviour is becoming a reliable tool to underpin scientific investigations and complement traditional theoretical and experimental approaches. In cases where an understanding of the dual nature of the structure of matter - continuous when viewed at large length scales and discrete when viewed at an atomic scale - and its interdependencies are crucial, multiscale materials modelling (MMM) approaches are required to complement continuum and atomistic analyses methods. At transitional (or microstructure) scales - *in between* continuum and atomistic - continuum approaches begin to break down, and atomistic methods reach inherent time and length-scale limitations (Ghoniem *et al.*, 2003). Transitional theoretical frameworks and modelling techniques are being developed to bridge the gap between length scale extremes. The power of analytical theories lies in their ability to reduce the complex collective behaviour of the basic ingredients of a solid (e.g. electrons, atoms, lattice defects, single crystal grains) into insightful relationships between cause and effect. For example, the description of deformation beyond the elastic regime is usually described by appropriate *constitutive equations*, and the implementation of such relationships within continuum mechanics generally relies on the inherent assumption that material properties vary continuously throughout the solid. However certain heterogeneities linked to either the microstructure or the deformation *per se* cannot be readily described within the framework provided by continuum mechanics: dislocation patterns, bifurcation phenomena, crack nucleation in fatigue, some non-local phenomena, etc. Some examples will be discussed next to illustrate the role of MMM in nano and micro-mechanics research.

Recent interest in *nanotechnology* is challenging the scientific community to analyze, develop and design nano to micro-meter size devices for applications in new generations of computers, electronics, photonics and drug delivery systems. These new exciting application areas require novel and sophisticated physically-based approaches for design and performance prediction. Thus theory and modelling are playing an ever increasing role in this area to reduce development costs and manufacturing times. An important problem which concerns the micro-electronic industry is the reliable operation of integrated circuits (ICs), where the lifetime is limited by the failure of interconnect wires in between sub-micron semiconducting chips. In some cases, the nucleation and growth of even a single nanovoid can cause interconnect failure. Statistical mechanics cannot adequately address this situation. Future electronic and optoelectronic devices are expected to be even smaller, with nanowires connecting nano-size memory and information storage and retrieval nano-structures. Understanding the mechanics of such nano-engineered devices will enable high levels of reliability and useful lifetimes to be achieved. Undoubtedly, defects are expected to play a major role in these nano and micro-systems due to the crucial impact on the physical and mechanical performance.

The potential of MMM approaches for computational materials design is also great. Such possibility was recently illustrated on a six-component Ti-base alloy, with a composition predetermined by electronic properties, which was shown to exhibit an entirely new twin- and dislocation-free deformation mechanism, leading to "superelasticity, superplasticity, superstrength, superworkability, Invar and Elinvar properties" (Saito *et al.*, 2003). Such an alloy would be unlikely to be found by trial and error. This points to a paradigm shift in modelling, away from reproducing known properties of known materials and towards simulating the behaviour of possible alloys as a forerunner to finding real materials with these properties.

In high-payoff, high-risk technologies such as the design of large-structures in the aerospace and nuclear industries, the effects of aging and environment on failure mechanisms cannot be left to conservative approaches. Increasing efforts are now focused on developing MMM approaches to develop new alloys and material systems in these areas. An illustration of an MMM based strategy for the development of large components surrounding the plasma core of a fusion energy system is shown in Figure 1 (Ghoniem *et al.*, 2003). The development of ultra-strong, yet ductile materials by combining nano-layers with different microstructures also requires detailed understanding of their mechanical properties. Such materials, if properly designed, may be candidates for many demanding applications (e.g. micro-electronics, opto-electronics, laser mirrors, aircraft structures, rocket engines, fuel cells, etc.).

Appropriate validation experiments are also crucial to verify that the models predict the correct behaviour at each length scale, ensuring that the linkages between approaches are directly enforced. However current nano and micro-scale mechanical experiments have been mostly limited to indentation (e.g.

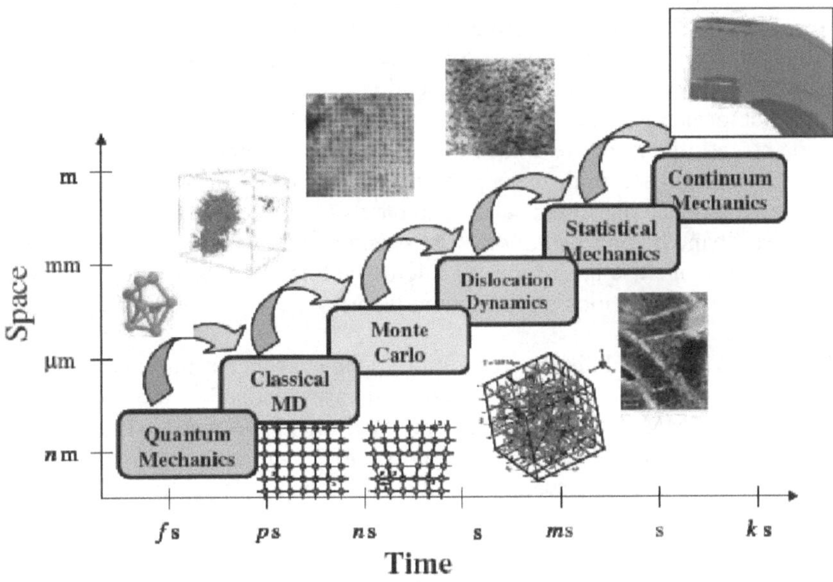

Figure 1. Illustration of a multiscale materials modelling approach for the design of radiation-resistant materials for fusion energy structures: MD, molecular dynamics.

(Tunvisut *et al.*, 2003) and bulge tests (e.g. Small *et al.*, 1994), and to non-contact tests such as X-ray residual stress measurements. Multi-scale interconnected approaches will need to be developed to interpret new and highly specialized nano/micro-mechanical tests. One of the advantages of these approaches is that, at each stage, physically meaningful parameters are predicted and used in subsequent models, avoiding the use of empiricism and fitting parameters.

As the material dimensions become smaller, its resistance to deformation is increasingly determined by internal or external discontinuities (e.g. surfaces, grain boundaries, dislocation cell walls, etc.). The Hall-Petch relationship has been widely used to explain grain size effects, although the basis of the relationship is strictly related to dislocation pileups at grain boundaries. Recent experimental observations on nano-crystalline materials with grains of the order of 10-20 nm indicate that the material is weaker than what would be expected from the Hall-Petch relationship (Campbell *et al.*, 1998). Thus, the interplay between interfacial or grain boundary effects and slip mechanisms within a single crystal grain may result in either strength or weakness, depending on their relative sizes. Although experimental observations of plastic deformation heterogeneities are not new, the significance of these observations has not been addressed till very recently. In same metallic alloys, regular patterns of highly localised deformation zones, surrounded by vast material volumes which contain little or no deformation, are frequently seen (e.g. Mughrabi *et al.*, 1987). The length scale associated with these patterns (e.g. typically the size of dislocation cells, the ladder spacing in persistent slip bands (PSB's), or the spacing between coarse shear bands) controls the material strength and ductility. As it may not be possible to homogenise such types of microstructures in an average sense using either atomistic simulations or continuum theories, new intermediate approaches will be needed.

The issues discussed above, in addition to the ever increasingly powerful and sophisticated computer hardware and software available, are driving the development of MMM approaches in nano and micro-mechanics. It is expected that within the next decade, new concepts, theories, and computational tools will be developed to make truly seamless multiscale modelling a reality.

2 Continuum Mechanics Methods

In this section, an overview will first be given of the main continuum mechanics-based framework used today to describe the non-linear deformation behaviour of materials at the local (e.g. single phase or grain level) and macroscopic (e.g. polycrystal level) scales. Emphasis will be placed on recent progress made in crystal plasticity, strain gradient plasticity, and homogenization techniques to link deformation phenomena simultaneously occurring at different scales in the material microstructure with its macroscopic behaviour.

Standard tensorial notation will be used throughout. Vectors will be described by boldface lower case letters, second order tensors by boldface upper case letters, and fourth order tensors by italic upper case letters. Also, $\boldsymbol{a}\cdot\boldsymbol{b} = \mathrm{a}_i b_j$, $\boldsymbol{Ab} = \mathrm{A}_{ij} b_j$, $\boldsymbol{A}:\boldsymbol{B} = \mathrm{A}_{ij} B_{ij}$, $\boldsymbol{AB} = \mathrm{A}_{ij} B_{jk}$, $\boldsymbol{L}:\boldsymbol{A} = \mathrm{L}_{ijkl} A_{kl}$, $(\boldsymbol{a}\otimes\boldsymbol{b})_{ij} = \mathrm{a}_i b_j$, where Einstein's summation applies for repeated indices.

2.1 Continuum Discretisation of a Boundary Value Problem

In a generic boundary value problem (BVP), the deformation of a body subjected to external forces and prescribed displacements is governed by the: (i) equilibrium equations, (ii) constitutive equations, (iii) boundary conditions, and (iv) initial conditions. The "weak" form of the boundary value problem is obtained when the equilibrium equations and the boundary conditions are combined into the "principle of virtual work". Such "weak form" constitutes the basis for obtaining a numerical solution of the deformation problem via, e.g. the finite element method. Thus, in a continuum mechanics Lagrangian formulation of a quasi-static BVP, the principle of virtual work is the vehicle by which the global equilibrium equations are obtained (see e.g. Zienkiewicz and Taylor (1994)).

The basic features of a generic Galerkin-type discretisation framework are given next.

Consider a structure occupying a domain V in the deformed configuration which is subjected to external forces and displacements on its boundary, Γ. In the absence of body forces and inertial effects, the principle of virtual work for the structure, in its rate form, satisfies the following equation,

$$\int_V \boldsymbol{\sigma} : \delta \dot{\in} \, dV - \int_\Gamma \mathrm{t} \cdot \delta \mathbf{v} \, d\Gamma = 0 , \quad (1)$$

for any arbitrary virtual velocity vector field $\delta\mathbf{v}$ compatible with all kinematics constraints. In the above equation, $\mathrm{t} = \boldsymbol{\sigma}\mathbf{n}$, represents the boundary traction forces, $\boldsymbol{\sigma}$ the Cauchy stress, $\mathbf{n}$ the normal to the surface on which the tractions act, and the virtual strain rate associated with the velocity field $\delta\mathbf{v}$.

To solve a complex BVP numerically, the discretisation of the principle of virtual work is generally performed using the finite element method. Let $\mathbf{v}$ be approximated at a material point within an element by,

$$\mathbf{v} = \sum_{i=1}^{N_{max}} N^i \hat{\mathbf{v}}^i \equiv \mathbf{N}\hat{\mathbf{v}} , \quad (2)$$

where $\hat{\mathbf{v}}$ denotes the nodal values of the element velocity field and $\mathbf{N}$ are the isoparametric shape functions. Substituting Eq. 2 into 1 leads to the discretised version of the principle of virtual work on the finite element V_e,

$$\mathbf{r}\{\hat{\mathbf{v}}\} \equiv \mathbf{f}^{\mathrm{int}} - \mathbf{f}^{\mathrm{ext}} = 0 , \quad (3)$$

where

$$\mathbf{f}^{\mathrm{int}} = \int_{V_e} \mathbf{B}^T \boldsymbol{\sigma} \, dV_e , \; \mathbf{f}^{\mathrm{ext}} = \int_{\Gamma_e} \mathbf{N}^T \mathbf{t} \, d\Gamma_e \quad (4)$$

are the internal and external global force vectors, respectively, and $\mathbf{B}$ relates the symmetric strain rate tensor with $\hat{\mathbf{v}}$. The global equilibrium relations (Eq. 3) represent a set of implicit non-linear equations which may be solved incrementally using a Newton-type algorithm. In a Newton-Raphson iterative scheme, the non-linear system (Eq. 3) is typically expanded using Taylor series in the neighbourhood of $\hat{\mathbf{v}}$,

$$\mathbf{r}\left\{\hat{\mathbf{v}}^k - \delta\hat{\mathbf{v}}^k\right\} = \mathbf{r}\left\{\hat{\mathbf{v}}^k\right\} + \frac{\partial \mathbf{r}\left\{\hat{\mathbf{v}}^k\right\}}{\partial \hat{\mathbf{v}}^k}\delta\hat{\mathbf{v}}^k + O\left\{\hat{\mathbf{v}}^{k2}\right\}, \tag{5}$$

where k represents a generic iteration and $\partial\mathbf{r}/\partial\hat{\mathbf{v}}$ is the global tangent stiffness or jacobian matrix of the non-linear system of equations. The formulation of accurate estimates of the global jacobian is at the heart of most numerical schemes developed to provide robust algorithms for the use of complex constitutive models with continuum approaches, e.g. see Esche, Kinzel and Altan (1997), Crisfield (1997), Busso, Meissonnier and O'Dowd (2000).

2.2 Continuum Approaches for Single Crystal Plasticity

Constitutive models developed to predict the anisotropic behaviour of single crystal materials generally follow either a Hill-type or a crystallographic approach. As a common feature, they treat the material as a continuum in order to describe properly plastic or visco-plastic effects. Hill-type approaches (e.g. Nouailhas, 1992, Schubert *et al.*, 2000) are based on a generalisation of the Mises yield criterion proposed by Hill (1950) to account for the non-smooth yield or flow potential surface required to describe the anisotropic flow stress behaviour of single crystals. In constitutive formulations based on crystallographic slip, the macroscopic stress state is resolved onto each slip system following Schmid's law. Internal state variables are generally introduced in both formulations to represent the evolution of the microstructural state during the deformation process. Although recent developments in these two approaches have now reached an advanced stage, the major improvements have been made by crystallographic models due to their ability to incorporate complex micro-mechanisms of slip within the flow and evolutionary equations of the single crystal models. These include the effects of dislocation interactions (e.g. Meric *et al.*, 1991), hardening and strain gradient phenomena (e.g. Gurtin, 2000), Meissonnier *et al.*, 2001), Stainer *et al.*, 2002), and general anisotropic plastic and visco-plastic behaviour (e.g. Busso and McClintock, 1996), Anand and Kothari, 1996). A brief outline of the salient features of local and non-local crystal plasticity approaches are given below.

2.3 Generic Local Crystallographic Framework

A generic internal variable based crystallographic framework is said to be a local one when the evolution of its internal variables can be fully determined by the local microstructural state at the material point. The description of the kinematics of most crystal plasticity theories follows that originally proposed in Asaro and Rice (1977), which has been widely reported in the computational mechanics literature (e.g. Pierce *et al.*, 1983; Kalidindi *et al.,* 1992; Busso *et al.*, 2000). It relies on the multiplicative decomposition of the total deformation gradient, **F**, into an inelastic, $\mathbf{F}^p$, and an elastic, $\mathbf{F}^e$, components. Thus, under isothermal conditions,

$$\mathbf{F} = \mathbf{F}^e \mathbf{F}^p . \tag{6}$$

Although single crystal laws can be formulated in a corotational frame, i.e. the stress evolution is computed on axes which rotate with the crystallographic lattice, the most widely used approach is to assume that the material's response is hyperelastic, that is its behaviour can be derived from a potential (i.e. free energy). Such potential may be expressed in terms of the elastic Green-Lagrange tensorial strain measure,

$$\mathbf{E}^e = \frac{1}{2}\left(\mathbf{F}^{e^{\mathrm{T}}} \mathbf{F}^p - 1\right), \tag{7}$$

and the corresponding objective work conjugate (symmetric) stress (Hill, 1975), or second Piola-Kirchhoff stress, **T**. Note that the Cauchy stress is related to **T** by

$$\boldsymbol{\sigma} = \det\left\{\mathbf{F}^e\right\}^{-1} \mathbf{F}^e \mathbf{T} \mathbf{F}^{e^{\mathrm{T}}} . \tag{8}$$

The hyperelastic response of the single crystal is governed by,

$$\mathbf{T} = \frac{\partial \Phi\left\{\mathbf{E}^e\right\}}{\partial \mathbf{E}^e}, \tag{9}$$

where $\partial\Phi/\partial\mathbf{E}^e$ represents the Helmholtz potential energy of the lattice per unit reference volume. Differentiation of Eq. 9, and assuming small elastic stretches, yields

$$\mathbf{T} \cong \mathbf{L} : \mathbf{E}^e . \tag{10}$$

where **L** is the anisotropic linear elastic moduli.

In rate-dependent formulations, the time rate of change of the inelastic deformation gradient $\mathbf{F}^p$, is related to the slipping rates on each slip system, $\dot{\gamma}^\alpha$ (Asaro and Rice, 1977), as

$$\dot{\mathbf{F}}^p = \left(\sum_{\alpha=1}^{n_\alpha} \dot{\gamma}^\alpha \mathbf{P}^\alpha \right) \mathbf{F}^p, \quad \text{with } \mathbf{P}^\alpha \equiv \mathbf{m}^\alpha \otimes \mathbf{n}^\alpha . \tag{11}$$

Here, $\mathbf{m}^\alpha$ and $\mathbf{n}^\alpha$ are unit vectors defining the slip direction and the slip plane normal on the slip system.

In rate-independent formulations, in contrast, flow rules are based on the well known Schmid law and a critical resolved shear stress, τ^α, whereby the rate of slip is related to the time rate of change of the resolved shear stress, $\tau^\alpha \left(= \mathbf{T} : \mathbf{P}^\alpha \right)$. Then,

$$\dot{\tau}^\alpha = \dot{\tau}_c^\alpha = \sum_{\alpha=1}^{n_\alpha} h^{\alpha\beta} \dot{\gamma}^\alpha , \quad \text{if } \dot{\gamma}^\alpha > 0 . \tag{12}$$

In the above equation, $h^{\alpha\beta}$, the slip hardening rates, incorporate latent hardening effects. Due to the severe restrictions placed on material properties, such as latent hardening, to ensure uniqueness in the mode of slip (e.g. Anand and Kothari, 1996; Busso and Cailletaud, 2005), and the associated difficulties in its numerical implementation, the use of rate-independent formulations has been somehow restricted and much more limited than rate-dependent ones. This has been compounded by the fact that, by calibrating their strain rate sensitivity response accordingly, rate-dependent models have been successfully used in quasi-rate-independent regimes. Thus, henceforth the focus of the discussions will be on rate-independent approaches.

The slip rate in Eq. 11 can functionally be expressed as,

$$\dot{\gamma}^\alpha = \hat{\dot{\gamma}}^\alpha \left\{ \tau^\alpha, S_1^\alpha, ..., S_{m_S}^\alpha, \theta \right\}, \tag{13}$$

where S_i^α (for $i = 1, ..., m_S$) denotes a set of internal state variables for the slip system α, and θ is the absolute temperature. A useful and generic expression for the overall flow stress in the slip system can be conveniently found by inverting Eq. 13 with $m_S = 2$,

$$\tau^\alpha = \pm \hat{f}_v^\alpha \left\{ \dot{\gamma}^\alpha, S_2^\alpha, \theta \right\} \pm c_S S_1^\alpha , \tag{14}$$

where c_S is a scaling parameter, and S_1^α and S_2^α represent additive and multiplicative slip resistances, respectively. Here the distinction between a multiplicative (S_1^α) and an additive (S_2^α) slip resistance is motivated by the additive and multiplicative use of non-directional hardening variables rather than on mechanistic considerations. By expressing the flow stress in the slip system in the way shown in Eq. 14, the contributions from viscous effects (first term in Eq. 14), and hardening mechanisms (second term) can be clearly identified. The majority of formulations relied on power law functions for Eq. 13, which results in $S_2^\alpha \neq 0$ and $S_1^\alpha = 0$ in Eq. 14 (e.g. Pierce *et al.*, 1983). This introduces a coupling between the viscous term and microstructure which is inconsistent with most strengthening mechanisms. Recently, work by Meric *et al.* (1991) and Busso *et al.* (2000) have proposed flow stress relations with $S_1^\alpha \neq 0$ and $S_2^\alpha = 0$, which allows a more physically meaningful interpretation of strengthening phenomena. For a more detailed discussions of these issues, see Busso and Cailletaud (2003).

The crystallographic formulation is completed with the evolutionary relations for the S_i^α internal slip system variables. The time rate of change of each internal slip system variable, $\dot{S}_i^\alpha$, is, in its most general form, expressed as,

$$\dot{S}_i^\alpha = \hat{\dot{S}}_i^\alpha \left\{ S_1^\alpha, \ldots, S_{m_S}^\alpha, \dot{\gamma}^1, \ldots, \dot{\gamma}^\alpha, \ldots, \dot{\gamma}^{n_\alpha}, \theta \right\}, \tag{15}$$

Note that the dependency of Eq. 15 on the slip rates on all systems enables cross-hardening effects to be accounted for.

2.4 Non-Local Approaches

The study of experimentally observed size-effects in a wide range of mechanics and materials problems has received a great deal of attention recently. Most continuum approaches and formulations dealing with these problems are based on strain-gradient concepts and are known as non-local theories since the material behaviour at a given material point depends not only on the local state but also on the deformation of neighbouring regions. Examples of such phenomena include particle size effects on composite behaviour (e.g. Nan *et al.*, 1996), precipitate phase size in two-phase single crystal materials (Busso *et al.*, 2000), increase in measured micro-hardness with decreasing indentor size (e.g. Swadener *et al.*, 2002), and decreasing film thickness (e.g. Huber *et al.*, 1999), amongst others.

The dependence of mechanical properties on length scales can in most cases be linked to features of either the microstructure, boundary conditions, or type of loading, which give rise to localised strain gradients. In general, the local material flow stress is controlled by the actual gradients of strain when the dominant geometric or microstructural length scales force the deformation to develop within regions of less than approximately 5 to 10 μm wide in polycrystalline materials, and of the order of 0.1 to 1 μm in single crystal materials (Busso *et al.*, 2000). Thus, gradient-dependent behaviour is expected to become important once the length scale associated with the local deformation gradients becomes sufficiently large when compared with the controlling microstructural feature (e.g. average grain size in polycrystal materials). In such cases, the conventional crystallographic formulations discussed in the previous section will be unable to predict properly the evolution of the local material flow stress. To accommodate these strain gradients, generation of geometrically necessary dislocations (GNDs) is required in these regions of incompatibility (e.g. Arsenlis and Parks, 2001; Busso *et al.*, 2000; Gao and Huang, 2003). The introduction of these GNDs, in addition to those stored in a random way (so-called "statistically stored" or SSDs), is what causes the additional strengthening of the material.

One of the first non-local theories was that proposed by Aifantis (1987), and Zbib and Aifantis (1988) to describe the formation of shear bands. This type of formulations rely on first and second derivatives of strain linked to the flow rule to describe strain gradient effects without the need to use higher order stresses. It requires additional boundary conditions, is relatively easy to implement numerically into the finite element method but is limited to describe strain-gradient problems that involve only one material length scale. Furthermore, by the nature of the formulation, it provides a limited mechanistic insight into the non-local phenomena.

A more physically intuitive continuum approach to describe strain gradient effects are constitutive theories such as those developed by Arsenlis and Parks (2001), Busso *et al.* (2000); Acharya (2000) and Bassani (2001). They rely on internal state variables to describe the evolution of the obstacle or dislocation network within the material and generally introduce the strain gradient effects directly in the evolutionary laws of the slip system internal variables without the need for higher order stresses. Thus, in some of these formulations, the functional dependency of the slip system internal variables evolutionary laws, such as the general form given for the slip resistance in Eq. 15, will now include an additional dependency in the gradient of the slip rates, $\nabla\dot{\gamma}^{\alpha}$. Then,

$$\dot{S}_i^{\alpha} = \hat{\dot{S}}_i^{\alpha}\left\{S_1^{\alpha},\ldots,S_{m_S}^{\alpha},\dot{\gamma}^1,\ldots,\dot{\gamma}^{\alpha},\ldots,\dot{\gamma}^{n_\alpha},\nabla\dot{\gamma}^1,\ldots,\nabla\dot{\gamma}^{\alpha},\ldots,\nabla\dot{\gamma}^{n_\alpha},\theta\right\}. \tag{16}$$

This class of theories has been shown capable of providing great physical insight into the effects of microstructure on the observed macroscopic phenomena, including rate-independent plastic deformation and visco-plasticity in both single crystal and polycrystalline materials (e.g. Arsenlis and Parks, 2001; Busso *et al.*, 2000; Acharya, 2000). One additional attractive aspect of the these theories is that they are relatively easy to implement numerically and do not require higher order stresses or additional boundary conditions. However, one limitation of these types of theories is that they are unable to describe problems which may require non-standard boundary conditions, such as the boundary layer problem modelled by Shu *et al.*, 2001.

A significant amount of work has been based on the treatment of the solid as a Cosserat continuum (e.g. Muhlhaus, 1989; Forest, 1998), where the material flow stress is assumed to be controlled not just by the rate of slip but also by the material curvature. However, even though Cosserat-type models have shown to be well suited to predict localisation phenomena, they are analytically complex and phenomenological in nature, and it is therefore difficult to gain a direct insight into the controlling physical phenomena at the microstructural level.

Another class of non-local theories with higher-order fields is that proposed recently by Fleck and Hutchinson (2001), and by Gurtin (2003). They contain higher order stresses and are extensions of the original theory proposed earlier by Fleck and Hutchinson (1997).

In such non-local formulation, the balance laws are based on a principle of virtual work where additional field variables, namely the work conjugates of the slip and slip rate gradients, are required. Consider a crystallographic approach and let **q** be a higher order stress vector, with components q_i, work conjugate to the slip rate gradients, $\nabla\dot{\gamma}^{\alpha}$, and π^{α} be the work conjugate to the slip rate, $\dot{\gamma}^{\alpha}$. Then,

$$\int_V \left[\boldsymbol{\sigma} : \delta\dot{\in} + \sum_{\alpha}\left(\pi^{\alpha} - \tau^{\alpha}\right)\delta\dot{\gamma}^{\alpha} + \sum_{\alpha}\mathbf{q}^{\alpha}\cdot\nabla\delta\dot{\gamma}^{\alpha}\right]dV$$

$$-\int_{\Gamma}\left[\boldsymbol{\sigma}\mathbf{n}\cdot\delta\mathbf{v} + \sum_{\alpha}\mathbf{q}^{\alpha}\cdot\mathbf{n}\delta\dot{\gamma}^{\alpha}\right]d\Gamma = 0. \tag{17}$$

By integrating Eq. 17 by parts, it can be shown that, in addition to the usual equilibrium relation in V,

$$div\,\boldsymbol{\sigma} = 0, \tag{18}$$

the following micro-force balance is obtained,

$$\pi^{\alpha} = \tau^{\alpha} + \nabla\mathbf{q} : \mathbf{1}, \tag{19}$$

where **1** is the second order unit tensor and τ^{α} is the resolved shear stress previously defined. Furthermore, Eq. 17 also implies that the following relations be satisfied at the boundary, Γ,

$$\mathbf{t} = \boldsymbol{\sigma}\mathbf{n} \quad \text{or} \quad \mathbf{v}, \quad \text{and} \quad \mathbf{q}^{\alpha}\cdot\mathbf{n}, \quad \text{or} \quad \dot{\gamma}^{\alpha} \tag{20}$$

It can be seen that, when no slip rate gradients are present, Eq. 17 gives $\pi^{\alpha} = \tau^{\alpha}$ and Eq. 17 resolves to the local version of the principle of virtual work given by Eq. 1.

The numerical implementation of these typically highly non-linear theories require the development of complex numerical algorithms. However, the introduction of higher order stresses and the required boundary conditions to satisfy the additional higher order field equations introduced by some non-local theories often precludes and limits the use of these formulations to in-house programmes (e.g. see implementation of a coupled stress theory by Shu *et al.*, 1999).

2.5 Homogenisation Approaches for Heterogeneous Microstructures

The bridging between the mechanical behaviour of heterogeneous materials and that of their individual constituents remains a topic of major interest and is at the heart of homogenisation schemes developed to predict the behaviour of materials at different scales. Such schemes are based on the assumption that the mechanical behaviour of individual constituents can lead to the description of the mechanical response of a macroscopic aggregate through either suitable interaction laws or a numerical averaging process.

When distinct heterogeneities exist at different microstructural levels, it is always possible to identify at each level the smallest possible representative volume element (RVE) of the microstructure which contains all the information concerning the distribution and morphology of the material's heterogeneities. Irrespective of the homogenisation schemes, once the relevant scales in the heterogeneous microstructure are identified, it is then necessary to select an RVE of the microstructure at the level of interest. Typically, the representative length scales in the microstructure are the average size of the largest heterogeneity at that particular level, D, and the size of the RVE, L. They must satisfy,

$$D << L . \tag{21}$$

The value of D can be, for instance, the average grain size in a polycrystal with randomly oriented grains of size, or be defined by the size of the precipitates relative to their mean spacing.

Homogenisation schemes require not only constitutive models for the individual constituents but also appropriate rules to make the transition between scales. The cases to be discussed here are those where loading in the RVE is homogeneous. If the loads applied on the RVE were inhomogeneous, then the homogenised equivalent medium is said to be a generalised one and special kinematics and equilibrium equations would apply (e.g. see Besson *et al.,* 2002). Some of the most widely used approaches to link local fields with macroscale phenomena, such as the large deformation and texture evolution of polycrystals, are Taylor-type (e.g. Kalidindi *et al.*, 1992) or Sachs-type (e.g. Leffers, 2001) models. The former assumes strain uniformity and can only fulfil compatibility at grain boundaries, but not equilibrium. In contrast, Sachs-type models assume homogeneous stresses and ignore local compatibility at grain boundaries. Amongst the most successful recent methods proposed to overcome the limitations of the assumed plastic strain or stress uniformity are those based on the relaxed constrained method (e.g. Van Houtte *et al.,* 1999) and on mean field approaches, such as self-consistent (e.g. Hill, 1965) and variational (e.g. Ponte Castaneda, (1991)) methods. Here, compatibility and equilibrium between grains are satisfied at both the local and the macroscopic levels.

In the self-consistent averaging approach, the interaction between a single crystal grain and its neighbours is treated as that between an inclusion with the same properties as those of the grain, embedded in an homogeneous equivalent medium (HEM) which has the same (unknown) properties as those of the macroscopic aggregate (see Fig. 2). A critical aspect of the self consistent method is that the strain distribution within the inclusion is assumed uniform when in reality, in cases of low strain rate sensitivity and large property mismatch, it is seldom the case.

One of the simplest self-consistent frameworks is that proposed by Hill (1965) where the stress and strain rate tensors in each phase or grain, $\dot{\mathbf{T}}$ and $\dot{\mathbf{E}}$, are related to those of the HEM, $\dot{\sigma}$ and $\dot{\varepsilon}$, through an elastic accommodation tensor $\mathbf{L}_e$,

$$\dot{\mathbf{T}} - \dot{\boldsymbol{\sigma}} = \mathbf{L}_e : \left(\dot{\in} - \dot{\mathbf{E}}\right). \tag{22}$$

The determination of suitable interaction relations between the inclusion and HEM is at the centre of most self-consistent approaches.

In Eq. 22, an elastic interaction between the grain and the polycrystal aggregate is implicitly assumed, thus a high constraint is imposed on the

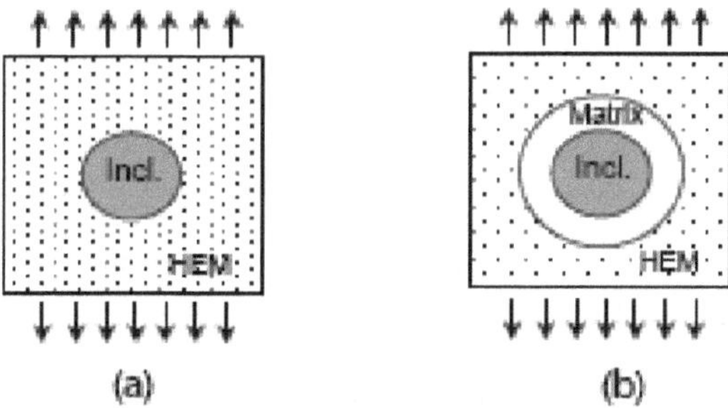

Figure 2. Comparison between the assumptions behind (a) standard and (c) generalised self-consistent approaches

inclusion by the surrounding elastic aggregate. In reality, such high constraint is partially relaxed by the plastic deformation of the polycrystalline aggregate. The work of Berveiller and Zaoui (1979) addressed this problem by introducing a plastic accommodation factor, $\mathbf{L}_p$. Thus, for the non-linear case, Eq. 22 becomes

$$\dot{\mathbf{T}} - \dot{\boldsymbol{\sigma}} = \mathbf{L}_p : \left(\dot{\in}^p - \dot{\mathbf{E}}^p\right). \tag{23}$$

where the global and local strain rate tensors are now the plastic ones. Similarly, the tangent approach method proposed by Molinari (2002) enables approximate solutions for non-linear material behaviour problems to be obtained while preserving the structure of the Eshelby's linear inclusion solution. Generally, self-consistent schemes are well suited to plastic or visco-plastic aggregates which can be treated as elastically rigid. The incorporation of elastic effects within a self-consistent framework is more difficult and clear solutions remain elusive. As self-consistent schemes are by definition implicit, their numerical implementation require Newton-type iterative procedures to solve the highly non-linear systems of equations. A more elaborate self-consistent approach (e.g. see Herve and Zaoui, 1993), referred to as a generalised or a three-phase scheme, is illustrated in Fig. 2b. Here, a composite sphere made up of two phases (i.e. matrix and inclusion) is embedded in an infinite matrix which is the HEM of unknown properties. The additional requirement in this case is that the average strain in the two-phase composite sphere must be the same as the strain prescribed in the far field. Herve and Zaoui (1993) showed that such scheme provides a framework for statistical analyses and demonstrated this by determining the effective behaviour of a random assembly of arbitrary-size spheres. Pitakthapanaphong and Busso (2002) used a similar generalised self-consistent approach to determine the elastoplastic properties of functionally gradient materials.

An alternative homogenisation method to the classical self-consistent approach is that derived from a variational procedure. The variational formulation proposed by Ponte Castaneda (1991) relies on the effective modulus tensor of linear elastic comparison composites, whereby the effective stress potentials of nonlinear composites are expressed in terms of the corresponding potentials for linear composites with similar microstructural distributions, to generate the corresponding bounds. Ponte Castaneda's variational principles have been used successfully to derive elastoplastic relations for metal matrix composites, amongst other applications. For further details, refer to references given herein.

As a result of the ever increasing computer power available, it is now becoming possible to replace the approximate mean field methods for more accurate ones based on numerical homogenisation. One of the most powerful is

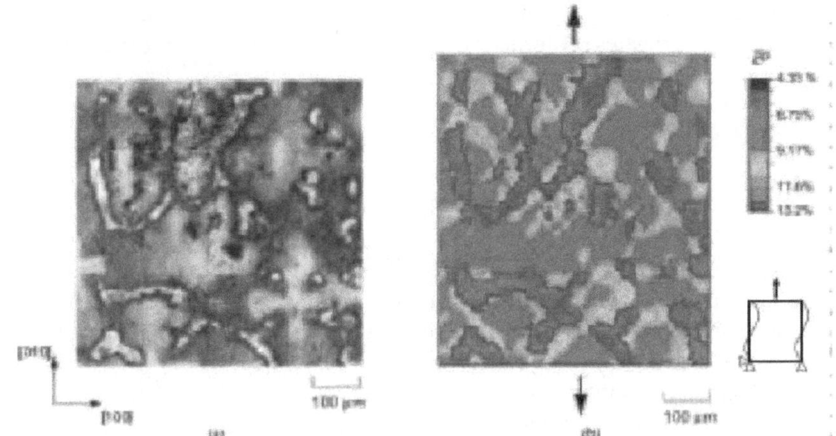

Figure 3. Homogenisation of a typical heterogeneous single crystal superalloy: (a) RVE and (b) predicted contours of accumulated inelastic strain under [010] uniaxial loading}

that based on periodic unit cell concepts, whereby a microstructural "window" of the constituents is periodically arranged and subjected to homogeneous far-field loading. By defining appropriate periodic boundary conditions on the smallest window or RVE which contains all the information about the local heterogeneities, the average stress and strain response of the unit cell can be obtained numerically. Even though these methods are computationally intensive, they offer the capability for digitised images of typical microstructures to be easily superimposed onto a regular FE mesh, thus enabling a precise model of the microstructure to be made and a framework to predict accurately the inhomogeneity of the deformation at the level of the individual phases. Alternatively, the FE mesh can be designed so that the element boundaries correspond to phase boundaries. This method has the advantage that almost any heterogeneity can be modelled, but the modelling effort required is generally impractical for very complex microstructures.

An example of a periodic unit cell of a typical heterogeneous single crystal superalloy RVE is presented in Fig. 3 (Regino *et al.*, 2002). Figure 3(a) shows an SEM micrograph of the microstructure's RVE at an intermediate scale, and (b) the predicted RVE contour plot of uniaxially equivalent accumulated inelastic strain, after a 10% straining along the [010] orientation at a rate of 10^{-3} 1/s at 950°C. It can be seen that the localisation of inelastic strain occurs in the vicinity of the stronger (eutectic) region. For complex 3D microstructures, such as those of polycrystal aggregates with randomly oriented grains, realistic FE meshes can be build based on Voronoi tessellations (e.g. see Ghosh *et al.*, 1996; Barbe *et al.*, 2002).

3 Outstanding Issues and Future Prospects

In this review, we have discussed the different modelling approaches which address specific phenomena at different length scales, and have highlighted the rich variety of physical, computational and technological issues within the broad area of nano and micromechanics which have been successfully addressed. We conclude by briefly summarising the current level of understanding in these areas, and discuss our expectations of forthcoming progress.

3.1 Bridging the Length and Time Scales

Even though recent advances in computing power have led to new and vastly improved simulation techniques at the atomistic and continuum levels, there has not yet been a concerted effort to develop explicit links between atomistic and continuum mechanics models. When properly reinforced, links between length scales should bring the overall field of material modelling one step closer to predict, *ab-initio*, the final properties of a proposed material.

A natural sequence to produce a fully integrated multi-scale modelling framework will require:

- The prediction of crystal energies, structures and the derivation of interatomic potentials from *ab-initio* calculations.
- The prediction of complex microstructural heterogeneities and morphologies resulting from solidification from the surface energies/kinetics predicted *ab-initio*.
- The identification of the crystal structures, slip systems and interface structures of the phases and the kinetics of phase transformations during subsequent thermo or thermo-mechanical processes using atomistic, dislocation dynamics and topological modelling techniques.
- The formulation of crystallographic models for each phase to describe the dominant deformation processes and the development of homogenisation schemes to obtain the macroscopic mechanical response of the material.

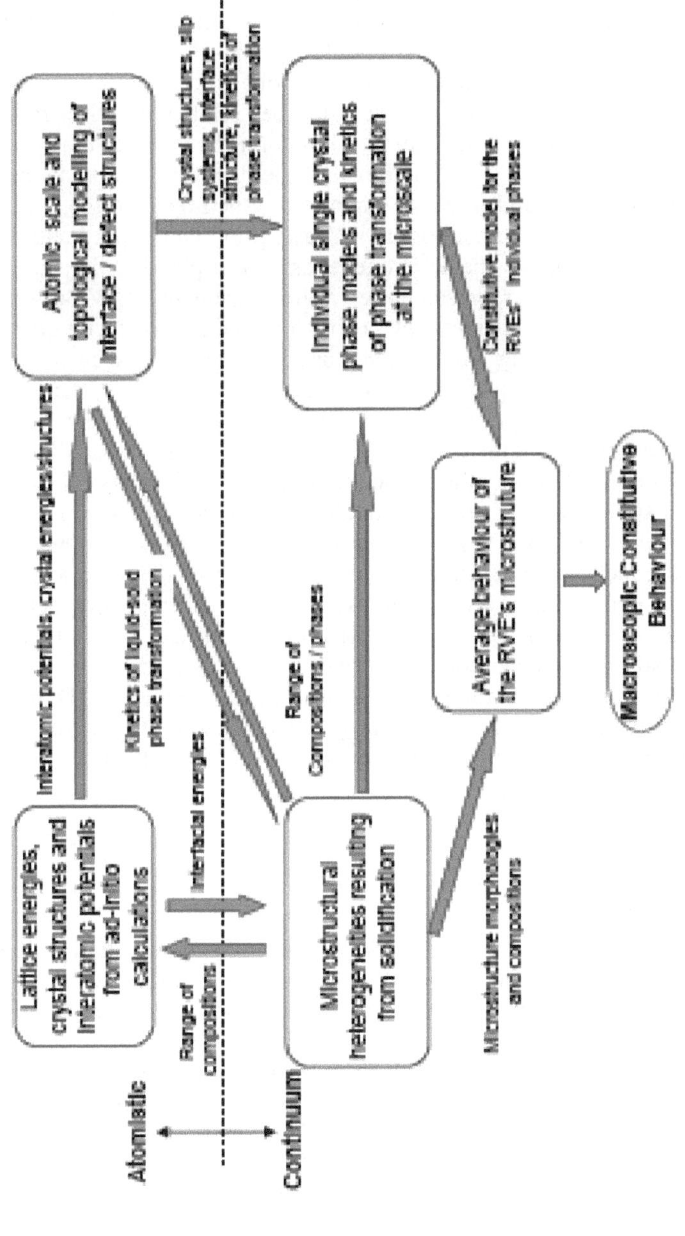

Figure 4. Flow chart showing the links between the different modelling approaches

Figure 4 illustrates the structure and the links between the modelling paradigms, together with the outputs and inputs for each sub-area. Here, the physical properties of the alloy are predicted at the atomistic scale and these are linked to the continuum level via appropriate constitutive and kinetic models. The properties determined by *ab-initio* calculations (e.g. crystal energies and structures, interatomic potentials) needs to be related to the range of compositions predicted after solidification. The *ab-initio* results can then be used in molecular dynamics (MD) calculations of deformation mechanisms. The resulting defect structures and interface kinetics are then fed into the continuum formulations to provide the crystallographic description and slip systems of the individual phases present in the heterogeneous (multi-phase) microstructures. The overall behaviour of representative volume elements (RVEs) of such microstructures can finally be determined using homogenisation techniques.

While progress on linking length scales has just started, linking time scales remains as an outstanding problem. As pointed out earlier, systematic and rigorous reduction of the degrees of freedom that describe material evolution will lead to self-consistent length scale linking, hence confident predictions of mechanical properties. On the other hand, methods for self-consistent linking of time scales are still lacking. Events at the atomic scale are often in the pico- to nano-second range, while microstructure evolution takes place on much longer time scales - seconds to years. Since the time evolution of the microstructure is path-dependent, events that occur in the pico- to nano-second time scale (e.g. atomic jumps, nucleation) may have profound effects on microstructure evolution. In addition, relaxation time scales in atomic models cannot be matched by continuum relaxation time scales. Thus, when an MD model is directly linked to the continuum, matching can be accomplished for static or quasi-static problems, while it has not been shown for fully-dynamic problems. Progress in this area is needed.

3.2 Atomistic Scales

Atomistic simulations have mainly supplemented experimentally obtained information till now. Nevertheless, future developments are expected in four major directions: (i) higher accuracy; (ii) larger systems; (iii)computationally faster methods, and (iv) more general approaches.

Density functional approaches have evolved rather rapidly during the past decade to address accuracy. The introduction of generalized gradient corrections has improved the accuracy of binding energies, surface energy and energy barriers. While the standard local spin density and generalized gradient approximations for the exchange-correlation energy can work for certain cases of strong correlation, i.e. where electrons partially preserve their localized atomic-like nature, they fail dramatically for others. Novel functionals, such as the recently developed self-correlation-free meta-GGA and self-interaction-free hyper-GGA, might yield a more reliable description of strong correlations. It is also possible that the use of quantum Monte Carlo methods for benchmark calculations could provide a fruitful path to assess and improve density functionals. Concurrent with the above efforts in developing more accurate functionals within the spirit of DFT, several other methods have been recently developed to treat the ground-state properties of condensed systems that exhibit strong correlations. These include the self-interaction correction method, the LDA+U method, the LDA+dynamical mean field theory method, and the Optimized Effective Potential method. In addition, the time-dependent extension of DFT and the GW approximation will provide a way to treat the excitation properties of moderately correlated electron systems.

Two other areas of developments, perhaps the most important ones, are faster methods, to enable dynamic processes to be simulated over sufficiently long time scales, and O(N) methods. Recasting the electronic structure calculations in an "order N" form which scales linearly with N will have important conceptual and practical implications for the treatment of sufficiently large systems. If the full calculation on a large system could be carried out in a time of "order N", then the properties of a small region of the large system could be calculated in a time independent of the system size. Such quantum "order N" methods will have exactly the same scaling as classical empirical potential methods. Furthermore the independence between the different regions will allow them to be readily adapted to parallel computations. This would enable calculations on large systems of great interest in areas which are beyond current capabilities, such as materials science and biology, to be performed.

A single method is unlikely to meet all goals in multiscale modelling of nano and micromechanics. Even with O(N) algorithms, it will not be possible in the foreseeable future to treat systems containing millions of atoms at a highly accurate DFT level using large basis sets, as would be necessary for certain materials science applications. Such problems can only be approached if one succeeds in linking methods of different accuracy, such as DFT methods with classical force fields, and applying the high-accuracy method only to regions where the low-accuracy method is expected to fail. Hybrid methods of this type will certainly be based on the same notions of locality as O(N) methods and will employ similar techniques.

In atomistic MD simulations, two areas are worth exploring. The first is the direct linking between atomistic and meso or microstructure-mechanics simulations. Approaches linking directly atomistic and continuum methods do exist, however a two-way connection between atomistic and dislocation dynamics simulations has not yet been achieved. Atomistic simulations are capable of providing interaction rules or mechanisms for a finite number of dislocation configurations, most of them being of high symmetry. In dislocation dynamics simulations, a few dislocations may interact at close proximity, forming a low symmetry configuration with respect to the simulation cell boundaries. It would be desirable to take such a configuration, and describe it in molecular dynamics simulations, and then feed the resulting configuration back to dislocation dynamics simulations in a seamless fashion. The second area is dislocation interaction with interfaces. A dislocation behaves differently near a surface or at a grain boundary, when compared to bulk dislocations. A fair understanding of how a dislocation nucleates at a surface or grain boundary (i.e. dislocation absorption or blockage at interfaces) exists. The interaction of a dislocation with an interface depends very much on the nature of the interface. Different grain boundary structures (e.g. low angle versus high angle grain boundaries) and compositions (e.g. pure vs. segregated grain boundaries) lead to a large number of interaction mechanisms. Many of these mechanisms remain to be explored, thus a comprehensive and simple description of dislocation-grain boundary interaction needs to be developed.

3.3 Transitional-Continuum Scales

The significant role that defects play in determining the mechanical properties of metallic materials and the vast progress recently made in computational techniques has led to the emergence of the field of Mesomechanics, which focuses on the behaviour of defects rather than those of atoms. Thus Mesomechanics describes the mechanics of heterogeneities and irregularities in materials, including topological defects (point, line, surface and volume), and compositional and structural heterogeneities. One of the most powerful mesomechanics methods is dislocation dynamics (DD), where considerable progress has been made during the past two decades due to a variety of conceptual and computational developments. It has moved from a curious proposal to a full-fledged and powerful computational method. In its present stage of development, DD has already addressed complex problems, and quantitative predictions have been validated experimentally. Progress in 3-D DD has contributed to a better understanding of the physical origins of plastic flow, and has provided tools capable of quantitatively describing experimental observations at the nano and micro-scales, such as the properties of thin films, nanolayered structures, micro-electronic components, and micro-mechanical elements. Advances in areas such as the mechanical behaviour of very small volumes and dislocations – interphase interactions are expected to continue in the near future.

One of the most difficult future challenges is the development of explicit links between the transitional-microstructure scales addressed by mesomechanics approaches and the continuum level. As previously discussed, mesomechanics approaches are needed to complement atomistic methods and provide information about defect interaction and the kinetics of slip and interphase motion. Such fundamental information can then be transferred to the continuum level to underpin the formulation of flow and evolutionary behaviour of continuum mechanics-based constitutive equations. Crystallographic approaches for single crystal behaviour which rely on internal slip system variables will continue to provide the most powerful framework to incorporate basic mechanistic understanding in continuum models. Numerical and analytical homogenisation techniques at the continuum level will be relied upon to a much greater extent than at present to model the behaviour of complex multi-phase and polycrystalline microstructures. This will enable the resulting constitutive models to incorporate explicit links between features of the microstructure at different levels and the macroscopic behaviour. However, further development of this type of multi-scale material design capability will require a few challenges to be overcome.

New and efficient computational techniques for processing and visualizing the enormous amount of data generated in mesomechanical and continuum multi-scale simulations must be developed. Then, the issue of computational efficiency must be addressed so that truly large-scale simulations on thousands of processors can be effectively performed. Another example of the severe computational complexities which can arise with deformation is given by the highly-intermittent nature of plastic slip, as it may require to be dynamically described as quick avalanche events separated by long time intervals, thus imposing severe computational limitations. Theoretical methodologies that enable the proper coarse graining of space should be pursued, such as multi-polar expansion techniques, continuum averaging of slip gradients, grains and heterogeneous microstructures, and extraction of average lattice curvatures. The increasing complexity of non-local formulations to predict size effects in multi-phase materials and composites will require improved and more robust numerical schemes to be developed, especially when a more physical description of dislocation interaction with themselves and with grain boundaries or other obstacles is required. Methods for a more direct coupling between DD simulations and continuum methods would also be required to improve the ability of non-local continuum mechanics approaches to predict complex deformation phenomena, such as size effects. For instance, details obtained by DD simulations can provide information about the evolution of the material topology, and the dislocation density tensor which can lead to precise descriptions of lattice curvatures.

The integration of the approaches discussed in this section is expected to lead to more physically-based multi-scale formulations and established materials-by-design as an approach to optimize the mechanical properties of materials with complex microstructures and to develop exciting new materials of extraordinary properties.

Bibliography

Acharya, A. and Beaudoin, A. (2000). Grain-size effect in viscoplastic polycrystals at moderate strains., *J. Mech. Phys. Solids* 48: 2213-2230.

Aifantis, E. (1987). The physics of plastic deformation., *Int. J. Plast.* 95: 211-247.

Anand, L. and Kothari, M. (1996). A computational procedure for rate-independent crystal plasticity, *J. Mech. Phys. Solids* 44: 525-558.

Arsenlis, A. and Parks, D. (2001). Modeling the evolution of crystallographic dislocation density in crystal plasticity, *J. Mech. Phys. Solids* 50: 1979-2009.

Asaro, R. J. and Rice, J. R. (1977). Strain localization in ductile single crystals., *J. Mech. Phys. Solids* 25: 309-338.

Barbe, F., Decker, l., Jeulin, D. and Cailletaud, G. (2002). Intergranular and intragranular behaviour of polycrytsalline aggregates. part i: FE model, *Int. J. Plast.* 17: 513.

Bassani, J. (2001). Incompatibility and a simple gradient theory of plasticity, *J. Mech. Phys. Solids* 49: 1983-1996.

Berveiller, B. and Zaoui, A. (1979). An extension of the self-consistent scheme to plastically owing polycrystals, *J. Mech. Phys. Solids* 26: 325-344.

Besson, J., Cailletaud, G., Chaboche, J.-L. and Forest, S. (2002). *Mécanique Non-Lineaire des Matériaux*, 1st edn, Hermes Science Publ., Paris.

Busso, E. and McClintock, F. (1996). A dislocation mechanics-based crystallographic model of a B2-type intermetallic alloy, *Int. J. Plasticity* 12: 1-28.

Busso, E. P., Meissonnier, F. T. and O'Dowd, N. P. (2000). Gradient-dependent deformation of two-phase single crystals, *J. Mech. Phys. Solids* 48: 2333-2361.

Busso, E.P. and Cailletaud, G. (2005), On the selection of active slip systems in crystal plasticity. *Int. J. Plasticity,* 21, 2212–2231.

Campbell, G., Foiles, S., Huang, H., Hughes, D., King, W., Lassila, D., Nikkel, D., Diaz de la Rubia, T., Shu, J. Y., and Smyshlyaev, V. P. (1998). Multi-scale modeling of polycrystal plasticity: a workshop report. *Mater. Sci. Engng,* A251, 1-22.

Crisfield, M. A. (1997). *Non-linear Finite Element Analysis of Solids and Structures*, Vol. 1 & 2, 4th edn, John Wiley & sons, New York.

Esche, S. K., Kinzel, G. L. and Altan, T. (1997). Issues in convergence improvement for non-linear finite element programs, *Int. J. Numer. Meth. Engng.* 40: 4577-4594.

Fleck, N. A. and Hutchinson, J. (1997). Strain gradient plasticity, in J. Hutchinson and T. Wu (eds), *Advances in Applied Mechanics*, Vol. 33, Academic Press, New York, pp. 295-361.

Fleck, N. and Hutchinson, J. (2001). A reformulation of strain gradient plasticity, *J. Mech. Phys. Solids* 49: 2245-2271.

Forest, S. (1998). Modelling slip, kink and shear banding in classical and generalised single crystal plasticity, *Acta Mater.* 46: 3265-3281.

Gao, H. and Huang, Y. (2003). Geometrically necessary dislocation and size-dependent plasticity, *Scripta Met.* 48: 113-118.

Ghoniem, N.M., Busso, E.P., Huang, H., and Kioussis, N. (2003) Multiscale modelling of nanomechanics and micromechanics: an overview. *Philosophical Magazine*, 83: 3475-3528.

Ghosh, S., Nowak, Z. and Lee, K. (2002). Quantitative characterisation and modelling of composite microstructures by voronoi cells, *Acta Mater.* 45: 2215.

Gurtin, M. E. (2000). On the plasticity of single crystals: free energy, microforces, plastic-strain gradients, *J. Mech. Phys. Solids* 48: 989-1036.

Gurtin, M. (2003). On a framework for small deformation visco-plasticity: free energy, microforces, strain gradients, Int. J. Plast. 19: 47-90.

Herve, E. and Zaoui, A. (1993). N-layered inclusion based micromechanical modelling, *Int. J. Eng. Sci.* 31: 1.

Hill, R. (1950). *The Mathematical Theory of Plasticity*, 4th edn, Clarendon Press, Oxford, U.K.

Hill, R. (1965). Continuum micromechanics of elasto-plastic polycrystals, *J. Mech. Phys. Solids* 14: 89-101.

Huber, N. and Tsakmakis, C. (1999). Determination of constitutive properties from spherical indentation data using neural networks. Part II: plasticity with nonlinear isotropic and kinematic hardening, *J. Mech. Phys. Solids* 47: 1589-1607.

Kalidindi, S., Bronkhorst, C. and Anand, L. (1992). Crystallographic texture theory in bulk deformation processing of fcc metals, *J. Mech. Phys. Solids* 40: 537.

Leffers, T. (2001). A model for rolling deformation with grain subdivisions. part i: the initial stage, *Acta Mat.* 17: 469.

Meissonnier, F., Busso, E. and O'Dowd, N. (2001). Finite element implementation of a generalised non-local rate-dependent crystallographic formulation for finite strains, *Int. J. Plasticity* 17: 601-640.

Meric, L., Poubanne, P. and Cailletaud, G. (1991). Single crystal modeling for structural calculations: part 1—model presentation, *J. Eng. Mat. Tech.* 162-170: 947-957.

Molinari, A. (2002). Averaging models for homogeneous viscoplastic and elastic visco-plastic materials, *J. Eng. Mat. Tech.* 124: 62.

Muhlhaus, H. B. (1989). Application of Cosserat theory in numerical solutions of limit load problems, *Int. Arch.* 59: 124-137.

Mughrabi, H. (1987). A two-parameter description of heterogeneous dislocation distributions in deformed metal crystals. *Mater. Sci. Engng*, 85, 15.

Nan, C.-W. and Clarke, D. (1996). The influence of particle size and particle fracture on the elastic-plastic deformation of metal matrix composites, Acta Mater. 44: 3801-3811.

Nouailhas, D. and Freed, A. (1992). A viscoplastic theory for anisotropic materials. *J. Eng. Mat. Tech.* 114: 97-104.

Pierce, D., Asaro, R. J. and Needleman, A. (1983). Material rate dependence and localized deformation in crystalline solids., *Acta Metall.* 31: 31951-1976.

Pitakthapanaphong, S. and Busso, E. (2002). Self-consistent elastoplastic stress solutions for functionally graded material systems subjected to thermal transients, *J. Mech. Phys. Solids* 50: 695-716.

Ponte Castaneda, P. (1991). The effective mechanical properties of nonlinear isotropic composites, *J. Mech. Phys. Solids* 39: 45-71.

Regino, M., Busso, E., O'Dowd, N. and Allen, D. (2002). Constitutive approach to model the mechanical behaviour of inhomogeneous single crystal superalloys., in L.-B. J. et al. (ed.), *Materials for Advanced Power Engineering*, Part I, Forschungszentrum Julich, GmbH, pp. 283-291.

Saito, T., Furuta, T., Hwang, J., Kuramoto, S., Nishino, K., Suzuki, N., Chen, R., Yamada, A., Ito, K., Seno, Y., Nonaka, N., Ikehata, H., Nagasako, N., Imamoto, C., Ikuhara, Y., and Sakuma, T. (2003). Multifunctional alloys obtained via a dislocation-free plastic deformation mechanism, *Science*, 300: 464-467.

Small, M., Daniels, B., Clemens, B., and Nix, W. (1994). The elastic biaxial modulus of Ag-Pd multilayered thin films measured using the bulge test, *J. Mater. Res.*, 9: 25-30.

Shu, J., King, W. E. and Fleck, N. A. (1999). Finite elements for materials with strain gradient effects, *Int. J. Numer. Meth. Engng*. 44: 373-391.

Shu, J., Fleck, N., Van der Giessen, E. and Needleman, A. (2001). Boundary layers in constrained plastic ow: comparison of nonlocal and discrete dislocation plasticity, *J. Mech. Phys. Solids* 49: 1361-1395.

Schubert, F., Fleury, G. and Steinhaus, T. (2000). Modelling of the mechanical behaviour of the SC Alloy CMSX-4 during thermomechanical loading, *Modelling Simul. Sci. Eng*. 8: 947-957.

Stainer, L., Cutino, A. and Ortiz, M. (2002). A micromechanical model of hardening, rate sensitivity and thermal softening in BCC single crystals. *J. Mech. Phys. Solids* 50: 1511-1545.

Swadener, J., Misra, A., Hoagland, R. and Nastasi, M. (2002). A mechanistic description of combined hardening and size effects, *Scripta Met*. 47: 343-348.

Tunvisut, K., N.P. O'Dowd and Busso, E.P. (2001). Use of scaling functions to determine mechanical properties of thin coatings from microindentation tests, *Int. J. Solids and Structures*, 38: 335-351.

Van Houtte, p., Delannay, L. and Samajdar, I. (1999). Prediction of cold rolling textures and strain heterogeneity of steel sheets by means of the lumel model, *Textures Microstruct.* 31: 109.

Zbib, H. and Aifantis, E. (1988). On the localisation and post-localisation behaviour of plastic deformation, *Res. Mechanica* 23: 261-305.

Zienkiewicz, O. C. and Taylor, R. L. (1994). *The Finite Element Method*, 4th edn, McGraw-Hill, New York.

Analytical and Numerical Methods for Modeling the Thermomechanical and Thermophysical Behavior of Microstructured Materials

Helmut J. Böhm, Dieter H. Pahr and Thomas Daxner
Institute of Lightweight Design and Structural Biomechanics,
Vienna University of Technology, Vienna, Austria

Abstract Basic application-related aspects of two important groups of approaches to continuum micromechanics of inhomogeneous materials are presented, viz., mean field schemes and methods based on discrete microstructures. Emphasis is put on handling both thermomechanical and thermal conduction problems.
On this basis some issues and applications of continuum micromechanics are discussed. They comprise incremental Mori–Tanaka methods for finite strains, modeling of the thermomechanical and thermal conduction behavior of diamond particle reinforced metal matrix composites, windowing estimates for the macroscopic linear responses of inhomogeneous media, and modeling of the mechanical behavior of cellular materials.

1 Introduction

Microstructured or heterogeneous materials play important roles in materials science and technology. This group of materials encompasses, among others, composites, polycrystalline materials, porous and cellular materials, functionally graded materials, concrete, wood, and bone. Their behavior can be studied at a number of length scales ranging from sub-atomic scales, where quantum mechanical methods must be used, to scales for which continuum descriptions are best suited. Continuum models are the topic of the present contribution.

In materials modeling transitions from lower to higher length scales aim at achieving a marked reduction of the number of degrees of freedom describing the system. Obviously, suitable choices of the degrees of freedom used at the higher length scale and of methods for carrying out the transition are preconditions for a successful bridging of the scales. The continuum methods discussed in the following are suitable for handling scale transitions from length scales in the low micrometer range to macroscopic samples, components or structures with sizes of millimeters to meters. The pertinent research field is usually referred to as continuum micromechanics of materials. The length scale of the inhomogeneities is termed the microscale, that of samples or components is the macroscale, and intermediate length scales are called mesoscales. A typical application of continuum micromechanics is studying composite materials in terms of the behavior and geometrical arrangement of their constituents, i.e., matrix and reinforcements such as fibers or particles.

Because the steady-state thermomechanical and thermal conduction behaviors of microstructured materials are of major practical importance, the present contribution concentrates on these two groups of problems. The mathematical descriptions of steady-state thermoelasticity and thermal conduction of heterogeneous materials share many common features and can be attacked using similar techniques. Table 1 lists the principal variables of the two groups of problems such that the analogies between them are highlighted. Other steady-state diffusion phenomena that are mathematically analogous to heat conduction are listed by Hashin (1983) and further problems are discussed by Torquato (2002).

Table 1. Principal variables in steady state elasticity and heat conduction problems.

physical problem	elasticity	thermal conduction
"direct variable"	displacement field $\mathbf{u}$ [m]	temperature field T [K]
generalized intensity	strain field $\boldsymbol{\varepsilon}$ [ ]	thermal gradient field $\mathbf{d}$ [Km^{-1}]
generalized flux	stress field $\boldsymbol{\sigma}$ [Pa]	heat flux field $\mathbf{q}$ [Wm^{-2}]
generalized property	elasticity $\mathbf{E}$ [Pa]	thermal conductivity $\mathcal{K}$ [$\mathrm{Wm}^{-1}\mathrm{K}^{-1}$]

An important difference between elasticity and conduction problems concerns the orders of the tensors involved, which is lower in the latter case. The displacements $\mathbf{u}$ are vectors whereas the temperatures T are scalars, stresses $\boldsymbol{\sigma}$ and strains $\boldsymbol{\varepsilon}$ are tensors of order 2, whereas the heat fluxes $\mathbf{q}$ and thermal gradients $\mathbf{d}$ are vectors, and the elasticity tensor $\mathbf{E}$ as well as its inverse, the compliance tensor $\mathbf{C} = \mathbf{E}^{-1}$, are of order 4, whereas the conductivity tensor $\mathcal{K}$ and its inverse, the resistivity tensor $\mathcal{R} = \mathcal{K}^{-1}$, are of order 2. The differences in the orders of the tensors directly affect the number of generalized moduli required for describing the generalized material property tensors as well as their symmetry properties (Nye, 1957). For example, cubic geometrical symmetry gives rise to macroscopic cubic symmetry in elasticity (with three independent elastic moduli) but isotropy in thermal conduction.

In the following, Nye notation is used for mechanical variables, i.e., tensors of order 4 are written as 6×6 quasi-matrices, and stress- as well as strain-like tensors of order 2 as 6-(quasi-)vectors. Conductivity-like tensors of order 2 are treated as 3×3 matrices. Tensors of order 4 are denoted by bold upper case letters, stress- and strain-like tensors of order 2 by bold lower case Greek letters, conductivity-like tensors of order 2 by calligraphic upper case letters, and 3-vectors by bold lower case letters. All other variables are taken to be scalars. The symbol "$*$" is used to denote the contraction of a tensor of order 2 and a 3-vector, i.e., $[\boldsymbol{\zeta} * \mathbf{z}]_i = \zeta_{ij} z_j$.

The bridging of length scales, which constitutes the central issue of continuum micromechanics, involves the handling of two main tasks. On the one hand, the behavior at some larger length scale must be estimated or bounded by using information from a smaller length scale, a problem known as homogenization. In the case of continuum micromechanics this involves finding a homogeneous "reference material" that is energetically equivalent to a given heterogeneous material, the main inputs being the geometrical arrangement and the material behaviors of the constituents at the microscale. On the other hand, the local responses at the smaller length scale must be deduced from the loading conditions (and, where appropriate, from the load histories) on the larger length scale. This task is referred to as localization. In many continuum micromechanical methods, homogenization is less demanding than localization because the local fields tend to show a marked dependence on details of the microgeometry, i.e., the local geometry of the constituents.

Homogenization relations usually take the form of volume averages, so that the homogenized value of a variable $f(\mathbf{x})$, $\langle f \rangle$, is given by

$$\langle f \rangle = \frac{1}{\Omega} \int_{\Omega} f(\mathbf{x}) \, \mathrm{d}\Omega \quad , \tag{1}$$

where $\mathbf{x}$ stands for the position vector and Ω for the integration volume, i.e., the inhomogeneous volume element to be studied. Homogenization must be based on volume elements of finite size that are as representative as possible of the phase arrangement of the microstructured material to be studied. For the mechanical behavior, the energetic equivalence between the behaviors on the micro- and macroscales can be expressed by the relation

$$\langle \boldsymbol{\sigma}^{\mathrm{T}} \boldsymbol{\varepsilon} \rangle = \frac{1}{\Omega} \int_{\Omega} \boldsymbol{\sigma}^{\mathrm{T}}(\mathbf{x})\, \boldsymbol{\varepsilon}(\mathbf{x})\, \mathrm{d}\Omega = \langle \boldsymbol{\sigma} \rangle^{\mathrm{T}} \langle \boldsymbol{\varepsilon} \rangle \quad , \tag{2}$$

(Hill, 1967), where the $\boldsymbol{\sigma}(\mathbf{x})$ are general statically admissible stress fields, the $\boldsymbol{\varepsilon}(\mathbf{x})$ are general kinematically admissible strain fields, and ^{T} indicates the transpose of a tensor. Equation (2) is known as Hill's macrohomogeneity condition or the Mandel–Hill condition.

In standard micromechanics approaches the micro- and macroscales are assumed to be sufficiently different so that, on the one hand, the fluctuating fields at the lower length scale influence the behavior at the higher length scale only via their volume averages. On the other hand, gradients of the macrofields as well as composition gradients at the higher length scale must not be significant at the lower one, so that the macroscopic fields appear to be locally constant. If the above conditions are not met or if the material is not statistically homogeneous, higher order homogenization schemes (Kouznetsova et al., 2002) or embedding techniques must be employed.

Continuum approaches to handling the homogenization and localization of microstructured materials can be classified into two main groups. On the one hand, local fields in the constituents may be described by their volume averages, the microgeometry being introduced via statistical descriptors. This modeling strategy gives rise to Mean Field Approaches (MFAs) and is discussed in chapter 2. On the other hand, the fields in discrete microgeometries that are characteristic in some sense of the actual microstructure can be evaluated at high resolution, typically by numerical methods. Such Discrete Microfield Approaches (DMAs) are the subject of chapter 3. Finally, chapter 4 is devoted to the discussion of some issues and applications of continuum micromechanical models.

For more extensive and in-depth treatments of many of the concepts and methods involved in continuum micromechanics see e.g., Mura (1987), Aboudi (1991), Nemat-Nasser and Hori (1993), Suquet (1997), Markov and Preziosi (2000), Bornert et al. (2001), Milton (2002), Torquato (2002), Qu and Cherkaoui (2006), and the references given therein. Additional information can be found in Hashin (1983), Zaoui (2002), and Böhm (2004a).

2 Mean Field Methods and Variational Bounds

This chapter — with the exception of sections 2.4 and 2.7 — aims at presenting relationships "in parallel" for thermoelasticity and thermal conduction. Only the most important equations are given; for more detailed treatments see, e.g., Böhm (2007) and the references given therein.

2.1 Basic Mean Field Relations

In this section linear thermoelastic and/or linear heat conduction behaviors are assumed at both the constituent and macroscopic levels, i.e.,

$$\begin{aligned} \boldsymbol{\varepsilon} &= \mathbf{C}\boldsymbol{\sigma} + \boldsymbol{\alpha}\Delta T & \qquad \mathbf{d} &= \mathcal{R}\mathbf{q} \\ \boldsymbol{\sigma} &= \mathbf{E}\boldsymbol{\varepsilon} + \boldsymbol{\vartheta}\Delta T & \qquad \mathbf{q} &= \mathcal{K}\mathbf{d} \quad , \end{aligned} \tag{3}$$

where $\boldsymbol{\alpha}$ and $\boldsymbol{\vartheta}$ denote the thermal expansion and specific thermal stress tensors with $\boldsymbol{\vartheta} = -\mathbf{E}\boldsymbol{\alpha}$, and ΔT is an applied spatially uniform temperature difference with respect to some stress-free reference temperature (note that ΔT only activates the thermal expansion behavior and is not connected to the temperature fields in thermal conduction).

Phase-wise constant fields are obtained by volume averaging over the phase volumes $\Omega^{(p)}$ according to eqn. (1), and the localization relations for phase $^{(p)}$ can be written as

$$\begin{aligned} \langle\boldsymbol{\varepsilon}\rangle^{(p)} &= \bar{\mathbf{A}}^{(p)}\langle\boldsymbol{\varepsilon}\rangle + \bar{\boldsymbol{\eta}}^{(p)}\Delta T & \qquad \langle\mathbf{d}\rangle^{(p)} &= \bar{\mathcal{A}}^{(p)}\langle\mathbf{d}\rangle \\ \langle\boldsymbol{\sigma}\rangle^{(p)} &= \bar{\mathbf{B}}^{(p)}\langle\boldsymbol{\sigma}\rangle + \bar{\boldsymbol{\beta}}^{(p)}\Delta T & \qquad \langle\mathbf{q}\rangle^{(p)} &= \bar{\mathcal{B}}^{(p)}\langle\mathbf{q}\rangle \quad . \end{aligned} \tag{4}$$

Here $\bar{\mathbf{A}}^{(p)}$ and $\bar{\mathbf{B}}^{(p)}$ are the phase averaged mechanical strain and stress concentration tensors, $\bar{\boldsymbol{\eta}}^{(p)}$ and $\bar{\boldsymbol{\beta}}^{(p)}$ are the phase averaged thermal strain and stress concentration tensors, and $\bar{\mathcal{A}}^{(p)}$ and $\bar{\mathcal{B}}^{(p)}$ are the phase averaged thermal gradient and flux concentration tensors, respectively. These concentration tensors fulfill the relations

$$\begin{aligned} &\sum_{(p)} \xi^{(p)}\bar{\mathbf{A}}^{(p)} = \mathbf{I} & \qquad &\sum_{(p)} \xi^{(p)}\bar{\mathcal{A}}^{(p)} = \mathcal{I} \\ &\sum_{(p)} \xi^{(p)}\bar{\mathbf{B}}^{(p)} = \mathbf{I} & \qquad &\sum_{(p)} \xi^{(p)}\bar{\mathcal{B}}^{(p)} = \mathcal{I} \\ &\sum_{(p)} \xi^{(p)}\bar{\boldsymbol{\eta}}^{(p)} = \sum_{(p)} \xi^{(p)}\bar{\boldsymbol{\beta}}^{(p)} = \mathbf{o} & & \quad , \end{aligned} \tag{5}$$

where $\xi^{(p)}$ stands for the volume fraction of phase $^{(p)}$, $\mathbf{I}$ is the symmetric unit tensor of order 4, $\mathcal{I}$ is the unit tensor order 2, and $\mathbf{o}$ is the strain-like null tensor. The sums run over all phases $^{(p)}$ of the heterogeneous material. Voids and other flaws that are present in the microstructure must be explicitly accounted for. If they are of comparable size to the microstructural features of primary interest they can be treated as separate phases (with $\mathbf{E}^{(p)} = \mathbf{0}$ and $\mathcal{K}^{(p)} = \mathcal{O}$ for voids), whereas they are best homogenized into an effective matrix behavior if they are much smaller.

The phase averaged concentration tensors together with the thermoelastic and conduction tensors of the phases allow to express the macroscopic (effective) thermoelastic and conduction tensors of a microstructured material as

$$\begin{aligned} \mathbf{E}^* &= \sum_{(p)} \xi^{(p)} \mathbf{E}^{(p)} \bar{\mathbf{A}}^{(p)} & \mathcal{K}^* &= \sum_{(p)} \xi^{(p)} \mathcal{K}^{(p)} \bar{\mathcal{A}}^{(p)} \\ \mathbf{C}^* &= \sum_{(p)} \xi^{(p)} \mathbf{C}^{(p)} \bar{\mathbf{B}}^{(p)} & \mathcal{R}^* &= \sum_{(p)} \xi^{(p)} \mathcal{R}^{(p)} \bar{\mathcal{B}}^{(p)} \\ \boldsymbol{\alpha}^* &= \sum_{(p)} \xi^{(p)} (\bar{\mathbf{B}}^{(p)})^{\mathrm{T}} \boldsymbol{\alpha}^{(p)} & & , \end{aligned} \tag{6}$$

respectively. The relation for $\boldsymbol{\alpha}^*$ is known as the Levin (1967) formula.

Within the mean field framework there are a number of relationships that link stress and strain as well as gradient and flux concentration tensors, among them

$$\begin{aligned} \bar{\mathbf{A}}^{(p)} &= \mathbf{C}^{(p)} \bar{\mathbf{B}}^{(p)} \mathbf{E}^* & \bar{\mathcal{A}}^{(p)} &= \mathcal{R}^{(p)} \bar{\mathcal{B}}^{(p)} \mathcal{K}^* \\ \bar{\mathbf{B}}^{(p)} &= \mathbf{E}^{(p)} \bar{\mathbf{A}}^{(p)} \mathbf{C}^* & \bar{\mathcal{B}}^{(p)} &= \mathcal{K}^{(p)} \bar{\mathcal{A}}^{(p)} \mathcal{R}^* \quad . \end{aligned} \tag{7}$$

In addition, there are connections between the mechanical and thermal concentration tensors. All mechanical and thermal concentration tensors of an n-phase material can be obtained from $n-1$ mechanical concentration and conduction tensors, respectively.

2.2 Dilute Inhomogeneities

When an ellipsoidal homogeneous inclusion, i.e., a region that is embedded in a matrix consisting of the same material, is subjected to a uniform stress-free strain, referred to as a transformation strain, $\boldsymbol{\varepsilon}_{\mathrm{t}}$, the resulting constrained strain, $\boldsymbol{\varepsilon}_{\mathrm{c}}$, is homogeneous, as was shown by Eshelby (1957). The constrained strain can be described in terms of the Eshelby tensor $\mathbf{S}^{(\mathrm{i,m})}$ as

$$\boldsymbol{\varepsilon}_{\mathrm{c}} = \mathbf{S}^{(\mathrm{i,m})} \boldsymbol{\varepsilon}_{\mathrm{t}} \quad . \tag{8}$$

By using the concept of equivalent homogeneous inclusions this result can be extended to inhomogeneous inclusions (inhomogeneities) under applied far-field strain or stress loads. The Eshelby tensor is of order 4 and depends on the elastic properties of the matrix and on the shape parameters of the inclusion. Expressions for $\mathbf{S}^{(\mathrm{i,m})}$ pertinent to inclusions $^{(\mathrm{i})}$ of general symmetry in isotropic and transversally isotropic matrices $^{(\mathrm{m})}$ were given, e.g., by Mura (1987) and Withers (1989), respectively. The equivalent of

$\mathbf{S}^{(\mathrm{i,m})}$ in thermal conduction is the "diffusion Eshelby tensor", $\mathcal{S}^{(\mathrm{i,m})}$. It is of order 2 and for isotropic matrices depends only on the shape parameters of the inhomogeneity, see, e.g., Hatta and Taya (1986) or Duan et al. (2006).

On the basis of the Eshelby tensor expressions for the mechanical inhomogeneity concentration tensors of dilute composites can be obtained as

$$\begin{aligned} \mathbf{A}_{\mathrm{dil}}^{(\mathrm{i})} &= \{\mathbf{I} + \mathbf{S}^{(\mathrm{i,m})}\mathbf{C}^{(\mathrm{m})}[\mathbf{E}^{(\mathrm{i})} - \mathbf{E}^{(\mathrm{m})}]\}^{-1} \\ \mathbf{B}_{\mathrm{dil}}^{(\mathrm{i})} &= \{\mathbf{I} + \mathbf{E}^{(\mathrm{m})}[\mathbf{I} - \mathbf{S}^{(\mathrm{i,m})}][\mathbf{C}^{(\mathrm{i})} - \mathbf{C}^{(\mathrm{m})}]\}^{-1} \end{aligned} \qquad (9)$$

(Hill, 1965a), provided the interfaces between the phases show ideal mechanical and thermal behavior. The thermal inhomogeneity concentration tensors can be expressed in analogy as

$$\begin{aligned} \mathcal{A}_{\mathrm{dil}}^{(\mathrm{i})} &= \{\mathcal{I} + \mathcal{S}^{(\mathrm{i,m})}\mathcal{R}^{(\mathrm{m})}[\mathcal{K}^{(\mathrm{i})} - \mathcal{K}^{(\mathrm{m})}]\}^{-1} \\ \mathcal{B}_{\mathrm{dil}}^{(\mathrm{i})} &= \{\mathcal{I} + \mathcal{K}^{(\mathrm{m})}[\mathcal{I} - \mathcal{S}^{(\mathrm{i,m})}][\mathcal{R}^{(\mathrm{i})} - \mathcal{R}^{(\mathrm{m})}]\}^{-1} \quad . \end{aligned} \qquad (10)$$

Equations (9) and (10) pertain to dilute ellipsoidal inhomogeneities that do not interact either directly or collectively, so that an individual inhomogeneity embedded in the matrix directly "feels" the far field stress, strain, gradient or flux. As a consequence the inhomogeneity concentration tensors are independent of the phase volume fractions. Equations (9) and (10) are excellent approximations for $\xi^{(\mathrm{i})} \lessapprox 0.01$ and usually do not lead to major errors for $\xi^{(\mathrm{i})} \lessapprox 0.1$.

The fields in the matrix are not homogeneous in the neighborhood of inclusions and inhomogeneities (Eshelby, 1959) and can be described in terms of the "exterior point Eshelby tensor", see, e.g., Ju and Sun (1999).

2.3 Non-Dilute Inhomogeneities

Within the Mean Field framework non-dilute inhomogeneity volume fractions are handled by describing interactions between inhomogeneities in a collective way rather than by accounting for interactions between individual inhomogeneities. There are two main strategies for doing so, effective field and effective medium approaches.

Mori–Tanaka Methods In effective field methods the perturbations "felt" by a given inhomogeneity due to the presence of all other inhomogeneities are approximated by a suitable average strain or stress in the matrix. This idea goes back to Brown and Stobbs (1971) as well as Mori and Tanaka (1973). Such methods, which are usually referred to as Mori–Tanaka methods (MTM), were proposed by a number of authors. A concise and flexible formulation was given by Benveniste (1987a), who expressed the Mori–Tanaka inhomogeneity concentration tensors in terms of the dilute concentration tensors as

$$\begin{aligned} \bar{\mathbf{A}}^{(i)}_{\mathrm{MT}} &= \mathbf{A}^{(i)}_{\mathrm{dil}} \bar{\mathbf{A}}^{(m)}_{\mathrm{MT}} & \bar{\mathcal{A}}^{(i)}_{\mathrm{MT}} &= \mathcal{A}^{(i)}_{\mathrm{dil}} \bar{\mathcal{A}}^{(m)}_{\mathrm{MT}} \\ \bar{\mathbf{B}}^{(i)}_{\mathrm{MT}} &= \mathbf{B}^{(i)}_{\mathrm{dil}} \bar{\mathbf{B}}^{(m)}_{\mathrm{MT}} & \bar{\mathcal{B}}^{(i)}_{\mathrm{MT}} &= \mathcal{B}^{(i)}_{\mathrm{dil}} \bar{\mathcal{B}}^{(m)}_{\mathrm{MT}} \quad . \end{aligned} \tag{11}$$

By using eqns. (5) the Mori–Tanaka matrix and gradient concentration tensors, in turn, can be obtained as

$$\begin{aligned} \bar{\mathbf{A}}^{(m)}_{\mathrm{MT}} &= \Big[\xi^{(m)}\mathbf{I} + \sum_{(i)\neq(m)} \xi^{(i)} \mathbf{A}^{(i)}_{\mathrm{dil}}\Big]^{-1} & \bar{\mathbf{A}}^{(i)}_{\mathrm{MT}} &= \mathbf{A}^{(i)}_{\mathrm{dil}} \Big[\xi^{(m)}\mathbf{I} + \sum_{(i)\neq(m)} \xi^{(i)} \mathbf{A}^{(i)}_{\mathrm{dil}}\Big]^{-1} \\ \bar{\mathcal{A}}^{(m)}_{\mathrm{MT}} &= \Big[\xi^{(m)}\mathcal{I} + \sum_{(i)\neq(m)} \xi^{(i)} \mathcal{A}^{(i)}_{\mathrm{dil}}\Big]^{-1} & \bar{\mathcal{A}}^{(i)}_{\mathrm{MT}} &= \mathcal{A}^{(i)}_{\mathrm{dil}} \Big[\xi^{(m)}\mathcal{I} + \sum_{(i)\neq(m)} \xi^{(i)} \mathcal{A}^{(i)}_{\mathrm{dil}}\Big]^{-1} \end{aligned} \tag{12}$$

respectively. The relations for the stress and flux concentration tensors are analogous.

By combining eqns. (6), (9), (11) and (12) explicit algorithms are obtained that support both homogenization and localization for materials that show a matrix–inclusion topology with aligned inhomogeneities. For the special case of two-phase materials consisting of spherical isotropic inhomogeneities in an isotropic matrix, simple scalar equations for the effective shear modulus G^*_{MT}, the effective bulk modulus B^*_{MT} and the effective conductivity K^*_{MT} can be obtained (Benveniste, 1987a; Hatta and Taya, 1986). For stiff inhomogeneities in a compliant matrix Mori–Tanaka methods underestimate the macroscopic stiffness, and for compliant inhomogeneities in a stiff matrix they overestimate it.

Self-Consistent Methods The idea underlying effective medium approaches is to approximate interaction effects between non-dilute inhomogeneities by embedding "phase patterns" into the homogeneous reference material rather than into the matrix and subjecting these "kernels" to the macroscopic fields. In its simplest form this concept uses a single inhomogeneity as the kernel and employs eqn. (6) to obtain the implicit relation

$$\mathbf{E}^*_{\mathrm{SC}} = \sum_{(p)} \mathbf{E}^{(p)} \mathbf{A}^{(p,*)}_{\mathrm{dil}} \qquad \mathcal{K}^*_{\mathrm{SC}} = \sum_{(p)} \mathcal{K}^{(p)} \mathcal{A}^{(p,*)}_{\mathrm{dil}} \quad , \tag{13}$$

where the superscript $^{(p,*)}$ indicates that the inhomogeneities $^{(p)}$ are embedded in the effective material of elasticity $\mathbf{E}^*_{\mathrm{SC}}$ or conductivity $\mathcal{K}^*_{\mathrm{SC}}$, so that

$\mathbf{A}_{\mathrm{dil}}^{(\mathrm{p},*)}$ is a function of $\mathbf{E}_{\mathrm{SC}}^*$ and $\bar{\mathcal{A}}^{(\mathrm{p},*)}$ is a function of $\mathcal{K}_{\mathrm{SC}}^*$, respectively. By plugging eqn. (10) into eqn. (13) and solving by self-consistent iteration the classical self-consistent scheme (CSCS) is obtained as

$$\begin{aligned}\mathbf{E}_{\mathrm{SC},n}^* &= \sum_{(\mathrm{p})} \mathbf{E}^{(\mathrm{p})} \{\mathbf{I} + \mathbf{S}_{n-1}^{(\mathrm{p},*)} \mathbf{C}_{\mathrm{SC},n-1}[\mathbf{E}^{(\mathrm{p})} - \mathbf{E}_{\mathrm{SC},n-1}]\}^{-1} \\ \mathbf{C}_{\mathrm{SC},n}^* &= (\mathbf{E}_{\mathrm{SC},n}^*)^{-1} \quad ,\end{aligned} \tag{14}$$

(Hill, 1965a). The CSCS expression for thermal conduction takes the form

$$\begin{aligned}\mathcal{K}_{\mathrm{SC},n}^* &= \sum_{(\mathrm{p})} \mathcal{K}^{(\mathrm{p})} \{\mathcal{I} + \mathcal{S}_{n-1}^{(\mathrm{p},*)} \mathcal{R}_{\mathrm{SC},n-1}[\mathcal{K}^{(\mathrm{p})} - \mathcal{K}_{\mathrm{SC},n-1}]\}^{-1} \\ \mathcal{R}_{\mathrm{SC},n}^* &= (\mathcal{K}_{\mathrm{SC},n}^*)^{-1}\end{aligned} \tag{15}$$

and relations analogous to the above equations can be given in terms of the stress and flux concentration tensors, respectively.

Equations (14) and (15) differ fundamentally from Mori–Tanaka expressions in being implicit and fully symmetric in terms of the phases (i.e., there is no distinction between matrix and inhomogeneities). Classical self-consistent methods pertain to materials that do not have a matrix–inclusion topology, such as polycrystals and composites with interwoven phases, aligned ellipsoidal grains being described in the former case. For the simplest case of two isotropic phases and macroscopic symmetry the CSCS gives rise to implicit equations for G_{SC}^*, B_{SC}^* and K_{SC}^*, the two former being coupled (Torquato, 2002).

Differential Schemes Another approach to constructing an effective medium theory takes the form of repeated cycles of adding small volume fractions of inhomogeneities followed by homogenization, increasingly larger inhomogeneities being added to the material (McLaughlin, 1977). For a two-phase material the resulting Differential Scheme (DS) in elasticity takes the form of the ordinary differential equations

$$\frac{\mathrm{d}\mathbf{E}_{\mathrm{D}}^*}{\mathrm{d}\xi^{(\mathrm{i})}} = \frac{1}{\xi^{(\mathrm{m})}}[\mathbf{E}^{(\mathrm{i})} - \mathbf{E}_{\mathrm{D}}^*]\mathbf{A}_{\mathrm{dil}}^{(\mathrm{i},*)} \tag{16}$$

(Hashin, 1988) and the analogous relationship for thermal conduction is given by

$$\frac{\mathrm{d}\mathcal{K}_{\mathrm{D}}^*}{\mathrm{d}\xi^{(\mathrm{i})}} = \frac{1}{\xi^{(\mathrm{m})}}[\mathcal{K}^{(\mathrm{i})} - \mathcal{K}_{\mathrm{D}}^*]\mathcal{A}_{\mathrm{dil}}^{(\mathrm{i},*)} \quad , \tag{17}$$

the initial conditions being

$$\mathbf{E}^*(\xi^{(\mathrm{i})} = 0) = \mathbf{E}^{(\mathrm{m})} \qquad \text{and} \qquad \mathcal{K}^*(\xi^{(\mathrm{i})} = 0) = \mathcal{K}^{(\mathrm{m})} \quad . \tag{18}$$

For isotropic spherical inhomogeneities in an isotropic matrix eqn. (16) simplifies to two coupled differential equations in $G^*_{\rm D}$ and $B^*_{\rm D}$, and eqn. (17) can be integrated to give a nonlinear equation in $K^*_{\rm D}$, see Torquato (2002). The differential scheme describes matrix–inclusion composites with aligned reinforcements that have a wide range of sizes.

2.4 Inelastic Inhomogeneous Materials

Mean Field Models for inhomogeneous materials with viscoelastic constituents are closely related to those for elastic composites. Relaxation moduli and creep compliances can be obtained by applying micromechanical approaches in the Laplace transformed domain, where the problems are analogous to elastic ones for the same microgeometries. Standard micromechanical methods can be used in conjunction with complex moduli for steady state vibration analysis of viscoelastic composites. For correspondence principles between descriptions pertaining to elastic and viscoelastic inhomogeneous materials see, e.g., Hashin (1983).

The extension of mean field estimates and bounding methods to elastoplastic, viscoelastoplastic, and damaged inhomogeneous materials has, however, proven to be challenging. The main difficulties in developing such methods lie, on the one hand, in representing the marked intra-phase fluctuations of the microstress and microstrain fields by their phase averages and, on the other hand, in the path dependence of plastic behavior.

In the literature several lines of development of mean field approaches for inhomogeneous materials with elastoplastic phases can be found. The most important of them have been secant plasticity concepts based on deformation theory, compare (Tandon and Weng, 1988) or (Ponte Castañeda and Suquet, 1998), and incremental plasticity models, see, e.g., Hill (1965b). Other relevant approaches are the tangent concept (Molinari et al., 1987) and affine formulations (Masson et al., 2000).

Secant models treat elastoplastic composites as nonlinear elastic materials and are, accordingly, limited to monotonic loading and radial (or approximately radial) trajectories of the constituents in stress space during loading. Advanced secant formulations, compare (Bornert and Suquet, 2001), are highly suitable for materials characterization, where they give excellent results.

Incremental mean field models are not subject to limitations with respect to that can be followed. They are based on formulations in terms of phase averaged instantaneous strain and stress rate tensors, $\mathrm{d}\langle\boldsymbol{\varepsilon}\rangle^{(\mathrm{p})}$ and $\mathrm{d}\langle\boldsymbol{\sigma}\rangle^{(\mathrm{p})}$, which are linked by instantaneous concentration tensors of the type

$$
\begin{aligned}
\mathrm{d}\langle\boldsymbol{\varepsilon}\rangle^{(\mathrm{p})} &= \bar{\mathbf{A}}_{\mathrm{t}}^{(\mathrm{p})}\mathrm{d}\langle\boldsymbol{\varepsilon}\rangle + \bar{\boldsymbol{\eta}}_{\mathrm{t}}^{(\mathrm{p})}\mathrm{d}T \\
\mathrm{d}\langle\boldsymbol{\sigma}\rangle^{(\mathrm{p})} &= \bar{\mathbf{B}}_{\mathrm{t}}^{(\mathrm{p})}\mathrm{d}\langle\boldsymbol{\sigma}\rangle + \bar{\boldsymbol{\beta}}_{\mathrm{t}}^{(\mathrm{p})}\mathrm{d}T \quad .
\end{aligned} \tag{19}
$$

Expressions for the instantaneous concentration tensors can be obtained in analogy to mean field methods in elasticity. For example, in the Incremental Mori–Tanaka (IMT) scheme of Pettermann (1997) the instantaneous matrix strain concentration tensor of a two-phase composite consisting of elastic inhomogeneities embedded in an elastoplastic matrix takes the form

$$
\bar{\mathbf{A}}_{\mathrm{t}}^{(\mathrm{p})} = \left\{\xi^{(\mathrm{m})}\mathbf{I} + \xi^{(\mathrm{i})}\left[\mathbf{I} + \mathbf{S}_{\mathrm{t}}^{(\mathrm{i,m})}\mathbf{C}_{\mathrm{t}}^{(\mathrm{m})}[\mathbf{E}^{(\mathrm{i})} - \mathbf{E}_{\mathrm{t}}^{(\mathrm{m})}]\right]^{-1}\right\}^{-1} \quad . \tag{20}
$$

Here $\mathbf{S}_{\mathrm{t}}^{(\mathrm{i,m})}$ is the instantaneous Eshelby tensor. It must be evaluated for the current, generally anisotropic, instantaneous stiffness tensor $\mathbf{E}_{\mathrm{t}}^{(\mathrm{m})}$, which typically has to be done numerically (Gavazzi and Lagoudas, 1990).

Incremental mean field schemes directly based on eqn. (20) tend to markedly overestimate the overall strain hardening in the post-yield regime, compare, e.g., Chaboche and Kanouté (2003). Recent developments involve the use of tangent operators that reflect the macroscopic symmetry of the composite, i.e., "isotropized" ones for statistically isotropic materials, see Bornert (2001), and algorithmic modifications, compare Doghri and Ouaar (2003) as well as Doghri and Friebel (2005). These improvements have succeeded in markedly reducing the tendency towards excessive hardening of incremental Mori–Tanaka methods for particle reinforced composites. Such modified IMT schemes are particularly attractive as micromechanically based material models that can be implemented at the integration point level into Finite Element programs for structural modeling. For a further discussion of aspects of IMT models see section 4.1.

2.5 Variational Bounds

Rigorous bounds for the overall thermomechanical and thermal conduction properties of inhomogeneous materials can be generated by combining variational principles with appropriate trial fields.

The simplest variational bounds are obtained from uniform stress, strain, gradient and flux fields together with minimum energy expressions (Voigt, 1889; Reuss, 1929; Wiener, 1912; Hill, 1952) in the form

$$\Big(\sum_{(\mathrm{p})} \xi^{(\mathrm{p})} \mathbf{C}^{(\mathrm{p})}\Big)^{-1} \leq \mathbf{E}^* \leq \sum_{(\mathrm{p})} \xi^{(\mathrm{p})} \mathbf{E}^{(\mathrm{p})}$$

$$\Big(\sum_{(\mathrm{p})} \xi^{(\mathrm{p})} \mathcal{R}^{(\mathrm{p})}\Big)^{-1} \leq \mathcal{K}^* \leq \sum_{(\mathrm{p})} \xi^{(\mathrm{p})} \mathcal{K}^{(\mathrm{p})} \quad . \tag{21}$$

These bounds describe the phase geometries by one-point statistics, i.e., by the phase volume fractions only, and hold for any volume element, but are typically too slack to be of practical use.

The variational formulation due to Hashin and Shtrikman corresponds to two-point statistics and, as a consequence, accounts for the macroscopic symmetry of phase arrangements. The original Hashin–Shtrikman bounds pertain to macroscopically isotropic inhomogeneous materials (Hashin and Shtrikman, 1962a,b) and analogous bounds were reported for a wide range of problems, among them the elastic behavior of composites reinforced by continuous fibers (Hashin and Rosen, 1964; Hashin, 1983), of materials with aligned ellipsoidal phases or anisotropic constituents (Willis, 1977), and of periodic media (Milton and Kohn, 1988). All mean field estimates discussed in section 2.4 fulfill the appropriate Hashin–Shtrikman bounds. For two-phase composites with aligned ellipsoidal reinforcements, in fact, the Mori–Tanaka estimates coincide with one of the Hashin–Shtrikman bounds (Weng, 1990) and the other can be obtained by exchanging the properties of matrix and inhomogeneities.

Bounds on the elastic and conduction behavior that are tighter than Hashin–Shtrikman bounds can be obtained by using three-point (or higher) statistics. Such "improved" bounds can be formulated to account for shapes and size distributions of inhomogeneities, which are introduced via microstructural parameters that are available for a range of important microgeometries (Torquato, 1998a, 2002).

The most flexible way of bounding the macroscopic elastoplastic behavior of heterogeneous media in a materials characterization context is due to Ponte Castañeda (1992), who introduced a variational principle that allows upper bounds on the effective nonlinear response to be generated on the basis of upper bounds for the elastic moduli.

2.6 Non-Aligned Inhomogeneities

The orientations of non-aligned inhomogeneities may be random (leading to isotropic macroscopic behavior), planar random (leading to transversally isotropic macroscopic behavior), or show general orientation distribution functions (ODFs), ρ. Within the mean field framework, the macroscopic properties of such materials can be estimated by orientation averaging over the elasticity tensor or over the mechanical concentration tensors. The former strategy and its variants essentially correspond to laminate analogy models (Fu and Lauke, 1998; Schjødt-Thomsen and Pyrz, 2001). The latter type of approach has been incorporated into Mori–Tanaka methods by a number of authors, the orientation averaging being done by direct numerical integration, see, e.g., Pettermann et al. (1997), or on the basis of expansions of the ODF in terms of generalized spherical harmonics (Viglin expansions), see, e.g., Advani and Tucker (1987) or Siegmund et al. (2004).

Following Duschlbauer et al. (2003) the starting point for such "extended" Mori–Tanaka methods are dilute inhomogeneity concentration tensors, $\mathbf{B}_{\rm dil}^{(\rm i)\angle}$ or $\mathcal{B}_{\rm dil}^{(\rm i)\angle}$, that have some prescribed orientation with respect to the global coordinate system. On the basis of eqn. (9) these concentration tensors can be expressed as

$$\begin{aligned}\mathbf{B}_{\rm dil}^{(\rm i)\angle} &= \mathbf{T}_\sigma^\angle \left\{\mathbf{I} + \mathbf{E}^{(\rm m)}[\mathbf{I} - \mathbf{S}^{(\rm i,m)}][\mathbf{C}^{(\rm i)} - \mathbf{C}^{(\rm m)}]\right\}^{-1} \mathbf{T}_\sigma^{\angle -1} \\ \mathcal{B}_{\rm dil}^{(\rm i)\angle} &= \mathcal{T}^\angle \left\{\mathcal{I} + \mathcal{K}^{(\rm m)}[\mathcal{I} - \mathcal{S}^{(\rm i,m)}][\mathcal{R}^{(\rm i)} - \mathcal{R}^{(\rm m)}]\right\}^{-1} \mathcal{T}^{\angle -1} \quad , \end{aligned} \qquad (22)$$

where $\mathbf{T}_\sigma^\angle$ stands for the stress transformation tensor from the local (reinforcement-based) to the global coordinate system and $\mathcal{T}^\angle$ for the corresponding conductivity-like transformation tensor of order 2. When Euler angles φ, ψ and θ are used, orientation averaging takes the form

$$\overline{\mathbf{B}}_{\rm dil}^{(\rm i)} = \langle\langle \mathbf{B}_{\rm dil}^{(\rm i)\angle} \rho \rangle\rangle = \int_0^{2\pi}\int_0^{2\pi}\int_0^{\pi} \mathbf{B}_{\rm dil}^{(\rm i)\angle}(\varphi,\psi,\theta)\, \rho(\varphi,\psi,\theta)\, {\rm d}\varphi\, {\rm d}\psi\, {\rm d}\theta \quad , \qquad (23)$$

the ODF being assumed to be normalized such that $\langle\langle \rho \rangle\rangle = 1$. Plugging $\overline{\mathbf{B}}_{\rm dil}^{(\rm i)}$ and its conduction equivalent, $\overline{\mathcal{B}}_{\rm dil}^{(\rm i)}$, into eqns. (11) and (12), respectively, orientation averaged Mori–Tanaka stress and flux concentration tensors are obtained as

$$\begin{aligned}\overline{\mathbf{B}}_{\rm MT}^{(\rm m)} &= [\xi^{(\rm m)}\mathbf{I} + \xi^{(\rm i)}\overline{\mathbf{B}}_{\rm dil}^{(\rm i)}]^{-1} & \overline{\mathcal{B}}_{\rm MT}^{(\rm m)} &= [\xi^{(\rm m)}\mathcal{I} + \xi^{(\rm i)}\overline{\mathcal{B}}_{\rm dil}^{(\rm i)}]^{-1} \\ \overline{\mathbf{B}}_{\rm MT}^{(\rm i)} &= \overline{\mathbf{B}}_{\rm dil}^{(\rm i)}\,\overline{\mathbf{B}}_{\rm MT}^{(\rm m)} & \overline{\mathcal{B}}_{\rm MT}^{(\rm i)} &= \overline{\mathcal{B}}_{\rm dil}^{(\rm i)}\,\overline{\mathcal{B}}_{\rm MT}^{(\rm m)} \quad , \end{aligned} \qquad (24)$$

which allow evaluating the effective thermoelastic and conductivity tensors from eqn. (6). Because $\overline{\mathbf{B}}_{\rm MT}^{(\rm i)}$ and $\overline{\mathcal{B}}_{\rm MT}^{(\rm i)}$ are orientation averages over all inhomogeneities, they are of limited use in assessing the stress state or heat flux in any given inhomogeneity. For such localization tasks the inhomogeneity stress and flux concentration tensors

$$\bar{\mathbf{B}}_{\rm MT}^{(\rm i)\angle} = \mathbf{B}_{\rm dil}^{(\rm i)\angle}\,\overline{\mathbf{B}}_{\rm MT}^{(\rm m)} \qquad \bar{\mathcal{B}}_{\rm MT}^{(\rm i)\angle} = \mathcal{B}_{\rm dil}^{(\rm i)\angle}\,\overline{\mathcal{B}}_{\rm MT}^{(\rm m)} \qquad (25)$$

may be used, which allow to evaluate the average stress state and flux in inhomogeneities of a given orientation (Duschlbauer, 2004). Analogous expressions can be formed in terms of the strain and gradient concentration tensors.

"Extended" Mori–Tanaka methods as described by eqns. (22) to (25) are not fully compatible with the Mori–Tanaka concept, which implicitly assumes aligned inhomogeneities (Ponte Castañeda and Willis, 1995). For this reason they can give rise to nonsymmetric effective "elastic tensors" in a number of scenarios, especially for multi-phase materials or when anisotropic phases are present (Benveniste et al., 1991; Ferrari, 1991). A mean field method that does not show this shortcoming was proposed by Ponte Castañeda and Willis (1995). Despite their ad-hoc nature extended Mori–Tanaka methods have been used with considerable success for studying the elastic behavior of short fiber reinforced composites and also of woven composites (Gommers et al., 1998).

2.7 Non-Ellipsoidal Inhomogeneities and Interfacial Thermal Conductances

The mean field methods for modeling the thermomechanical and conduction behavior of microstructured materials presented in sections 2.1 to 2.6 do not possess an absolute length scale. It is well known, however, that interfacial effects introduce an absolute length scale into the macroscopic thermal conductivity of inhomogeneous materials. The present section aims at presenting mean field models that can describe this behavior and/or the effects of non-ellipsoidal inhomogeneities.

In general, both finite interfacial thermal conductances and non-ellipsoidal shapes of inclusions or inhomogeneities cause the the Eshelby property, to be lost, i.e., the fields in the dilute inclusions become inhomogeneous. In such cases approximate solutions can be obtained by volume averaging the Eshelby tensors or the dilute inhomogeneity concentration tensors (these two strategies are not equivalent, eqns. (9) and (10) being nonlinear). Because volume averaged dilute inhomogeneity concentration tensors have the direct interpretation of describing appropriate ellipsoidal "replacement inhomogeneities" that have perfect interfaces (Duschlbauer, 2004), this modeling strategy is discussed in the following.

The resulting "replacement tensor" (RT) algorithm for two-phase materials uses numerical methods for computing the local fields in volume elements that contain a single inhomogeneity at a dilute volume fraction of, say, $\xi^{(i)}_{\rm dil} \lessapprox 0.001$ under six linearly independent mechanical and/or three linearly independent thermal load cases (Nogales, 2008). From these solutions dilute effective elasticity and conductivity tensors, $\mathbf{E}^*_{\rm dil}$ and $\mathcal{K}^*_{\rm dil}$, as well as dilute replacement inhomogeneity strain and gradient concentration tensors, $\mathbf{A}^{(\rm i,r)}_{\rm dil}$ and $\mathcal{A}^{(\rm i,r)}_{\rm dil}$, respectively, can be extracted. In the case of finite interfacial conductances the temperature jumps at the interfaces must be included into the replacement inhomogeneity gradient concentration tensor. This may be achieved, e.g., by evaluating the averaged matrix gradient concentration tensor and using the two-phase version of eqn. (5) for obtaining $\mathbf{A}^{(\rm i,r)}_{\rm dil}$ and $\mathcal{A}^{(\rm i,r)}_{\rm dil}$. Equations (6) provide consistency conditions

$$\begin{aligned}\mathbf{E}^{(\rm i,r)} &= \mathbf{E}^{(\rm m)} + \frac{1}{\xi^{(\rm i)}_{\rm dil}}[\mathbf{E}^*_{\rm dil} - \mathbf{E}^{(\rm m)}](\mathbf{A}^{(\rm i,r)}_{\rm dil})^{-1} \\ \mathcal{K}^{(\rm i,r)} &= \mathcal{K}^{(\rm m)} + \frac{1}{\xi^{(\rm i)}_{\rm dil}}[\mathcal{K}^*_{\rm dil} - \mathcal{K}^{(\rm m)}](\mathcal{A}^{(\rm i,r)}_{\rm dil})^{-1} \end{aligned} \tag{26}$$

from which the replacement inhomogeneity elasticity and conduction tensors $\mathbf{E}^{(\rm i,r)}$ and $\mathcal{K}^{(\rm i,r)}$ can be extracted. The replacement tensors $\mathbf{E}^{(\rm i,r)}$, $\mathbf{A}^{(\rm i,r)}_{\rm dil}$, $\mathcal{K}^{(\rm i,r)}$ and $\mathcal{A}^{(\rm i,r)}_{\rm dil}$ can then be inserted into mean field relations in lieu of the corresponding "standard" tensors $\mathbf{E}^{(\rm i)}$, $\mathbf{A}^{(\rm i)}_{\rm dil}$, $\mathcal{K}^{(\rm i)}$ and $\mathcal{A}^{(\rm i)}_{\rm dil}$, respectively. This strategy is well suited for use with Mori–Tanaka methods, where only one dilute configuration must be evaluated. The resulting MTM/RT models can handle, e.g., polyhedral particles with inhomogeneous interfacial conductances, compare section 4.2. Because the Mori–Tanaka method by default treats inhomogeneities as aligned, directional averaging may be required to achieve macroscopic isotropy in the case of anisotropic replacement inhomogeneities that are randomly oriented.

For inhomogeneities that have simple shapes and interfaces with homogeneous conductances the replacement inhomogeneity conductivity and gradient concentration tensors are diagonal tensors. For example, in the case of spherical particles with homogeneous interfacial conductances the two tensors are isotropic and their components are

$$K^{(\rm i,r)} = K^{(\rm i)}\frac{dh}{dh + 2K^{(\rm i)}} \qquad\qquad A^{(\rm i,r)}_{\rm dil} = \frac{3K^{(\rm m)}}{K^{(\rm m)} + K^{(\rm i,r)}} \quad , \tag{27}$$

(Benveniste and Miloh, 1986; Böhm and Nogales, 2008), where d stands for the particle diameter and h for the interfacial conductance. By combining these results with eqns. (11), (12) and (6) the well-known scalar expressions for the macroscopic conductivity of composites containing spherical particles with finite interfacial conductances given by Hasselman and Johnson (1987) and Benveniste (1987b) are recovered.

Approaches that are related to the replacement inhomogeneity concept can also be used to study the mechanical behavior of composites that are subject to damage by decohesion between reinforcement and matrix or by reinforcement failure, see, e.g., Sun et al. (2003).

An alternative mean field approach to studying materials containing non-ellipsoidal inhomogeneities was proposed by Kachanov et al. (1994). It is based on decomposing the elastic strain–stress relation of an elastic microstructured material as

$$\langle \boldsymbol{\varepsilon} \rangle^* = \mathbf{C}^* \langle \boldsymbol{\sigma} \rangle^* = [\mathbf{C}^{(\mathrm{m})} + \mathbf{C}_{\mathrm{c}}^{(\mathrm{i,m})}] \langle \boldsymbol{\sigma} \rangle^* \quad , \tag{28}$$

the tensor $\mathbf{C}_{\mathrm{c}}^{(\mathrm{i,m})} = \mathbf{C}^* - \mathbf{C}^{(\mathrm{m})}$ being referred to as the compliance contribution tensor. Dilute compliance contribution tensors can be evaluated from numerical analysis, and Mori–Tanaka or self-consistent schemes can be set up to handle non-dilute cases on this basis. This approach can also be extended to thermal conduction problems.

3 Methods Based on Discrete Microstructures

The most important micromechanical approaches based on discrete microstructures encompass unit cell, windowing and embedding methods, the emphasis in the present chapter being put on the former two. In addition, small microstructured samples may also be modeled in their entirety, usually with boundary conditions that correspond to some experimental situation, see e.g. (Papka and Kyriakides, 1999; Luxner, 2006). Depending on the boundary conditions applied, such models may be related to windowing methods.

3.1 General Remarks

Volume Elements Heterogeneous volume elements used in discrete microstructure models can range from highly idealized periodic geometries, such as simple cubic arrays of spheres in a matrix, to complex microgeometries that aim at capturing the statistics of the phase arrangements of real materials. Geometrically complex volume elements can be obtained either by computer simulation or by experimental techniques such as computed tomography (Buffière et al., 2008) or serial sectioning (Chawla and Chawla, 2006). Obviously, volume elements based on real microstructures in general are not periodic, whereas computer generated phase arrangements can be either periodic or non-periodic.

Algorithms for generating generic random microgeometrical models with matrix–inclusion microtopology typically involve random addition methods (also known as random sequential insertion models). They use a random process to select possible positions for new inhomogeneities, these "candidates" being accepted if they do not collide with any of the existing inhomogeneities and rejected otherwise. Such "hard core" models tend to be limited to moderate reinforcement volume fractions due to jamming and their representativeness for phase arrangements generated by mixing processes may be open to question (Stroeven et al., 2004). Marked improvements on both counts can be obtained, on the one hand, by using RSA geometries as starting configurations for random perturbation models that apply small random displacements to each inhomogeneity to find acceptable arrangements of individual reinforcements as the size of the volume element is reduced (Segurado, 2004). On the other hand, heuristic "stirring" models can be used to generate matrix-rich regions that support the placement of further inhomogeneities (Melro et al., 2008). Another strategy for producing volume elements, which is not limited to matrix–inclusion topologies, modifies starting configurations by simulated annealing and related procedures to generate "statistically reconstructed" phase arrangements that closely approach the phase distribution statistics of the target material (Rintoul and Torquato, 1997; Zeman, 2003). A further possible approach to the computer generation of volume elements for inhomogeneous materials involves simulation of the relevant production processes. For a discussion of many aspects of the computer generation of random microstructures see Jeulin (2001).

Computer generated microgeometries for composites have tended to employ idealized reinforcement shapes, equiaxed particles embedded in a matrix, for example, being typically represented by spheres, and fibers by cylinders or prolate spheroids. Real structure phase arrangements, in contrast, may involve very irregular particle shapes (Chawla and Chawla, 2006).

Size of Volume Elements For any discrete microstructure model the question immediately arises what level of geometrical complexity (and thus what size of volume element) is required for adequately representing the physical behavior of the inhomogeneous material to be studied. Representative Volume Elements (RVEs), which by definition are sufficiently large to make predictions independent of details of the selection of the geometry and of boundary conditions, provide a theoretical solution to the above problem. In practice, however, limitations in computer power have tended to restrict simulations to volume elements of rather limited size, which typically are only approximations to proper RVEs and have been referred to as Statistical Volume Elements or Sub-Representative Volume Elements (SVEs).

There are two main approaches to assessing the adequacy of the sizes of such volume elements. One of them is based on using statistical descriptors of the microgeometry, corresponding to the concept of "geometrical RVEs".

In this context adequate sizes of model geometries are estimated on the basis of purely geometrical parameters of the phase arrangement. This can be done on the basis of experimentally obtained correlation lengths (Bulsara et al., 1999; Jeulin, 2001) or by comparing statistical distribution functions of actual and model microgeometries (Rintoul and Torquato, 1997; Zeman and Šejnoha, 2001). The other approach assesses the dependence of the predictions for some macroscopic material behavior on the size of volume elements in analogy to the concept of "physical RVEs". This idea goes back to Hill (1963), who defined RVEs as giving the same macroscopic response irrespective of the boundary conditions used. For example, hierarchies of bounds from windowing analysis, compare section 3.3, differences between predictions for some macroscopic modulus or response (e.g., for homogenized stress vs. strain curves), or deviations from some macroscopic material symmetry have been employed for checking if a given volume element approaches being an RVE according to this concept. Criteria for assessing the sizes of model geometries were discussed and compared, e.g., by Trias et al. (2006) and Swaminathan et al. (2006).

Methods based on geometrical parameters or on the overall elastic behavior typically give rise to relatively small volume elements. For example, Zeman (2003) reported that two-dimensional volume elements for studying the transverse behavior of continuously reinforced composites only require 10 to 20 randomly positioned fibers. For the case of statistically isotropic elastic composites with matrix–inclusion topology and sphere-like particles of equal size Drugan and Willis (1996) estimated that for approximating the overall moduli with errors of less than 5%, non-periodic volume elements with sizes of approximately five particle diameters are sufficient for any volume fraction. The concept of physical RVEs obviously implies that adequate sizes of model geometries depend on the physical property to be studied. A number of numerical studies (Jiang et al., 2001; Gitman et al., 2006) have indicated that inelastic constituent behavior tends to lead to a requirement for larger volume elements for satisfactorily approximating the overall symmetries and for obtaining good agreement — especially at elevated strains — between the responses of phase arrangements designed to be statistically equivalent. For cases involving damage at the constituent level, in fact, eliminating the dependence of the homogenized responses on the size of the volume element may be impossible (Gitman et al., 2007).

Model microgeometries of similar size that describe the same material may be viewed as being (approximate) realizations of the same statistical process. As a consequence, ensemble averaging can be used to obtain improved estimates for the macroscopic behavior when predictions for a set of such volume elements are available (Kanit et al., 2003; Stroeven et al.,

2004). The number of different volume elements required for a given accuracy of the ensemble averages tends to decrease as the sizes of the models increase (Khisaeva and Ostoja-Starzewski, 2006).

Numerical Methods The majority of published micromechanical studies of discrete microgeometries have employed standard numerical engineering methods for resolving the microfields, work using Finite Difference and Finite Volume algorithms, Fast Fourier Transforms, the Boundary Element Method as well as the Finite Finite Element Method (FEM) having been reported. In addition, some specialized approaches such as Transformation Field Analysis (Dvorak, 1992; Michel and Suquet, 2003) as well as the the Method of Cells and its developments (Aboudi, 1996, 2004) have been used. For discretizing numerical methods the characteristic length of the discretization ("mesh size") must be considerably smaller than the microscale of a given problem in order to obtain spatially well resolved results.

At present, the Finite Element Method is the most commonly used numerical scheme for evaluating the fields in discrete microgeometries, especially in the nonlinear range, where its flexibility, efficiency and capability of supporting a wide range of constitutive models for the constituents and for the interfaces between them are especially appreciated. Another asset of the FEM in the context of continuum micromechanics is its ability to handle discontinuities in stress, strain, thermal gradient and heat flux components, which typically occur at interfaces between different constituents, in a natural way via appropriately placed element boundaries. In addition, phase averages of the local fields can be evaluated in a rather straightforward way by making use of the fact that in displacement based finite elements stresses and strains are given at the integration points (as are thermal gradients and heat fluxes in temperature-based heat conduction codes). This allows to approximate volume integrals as weighted sums of the type

$$\langle f \rangle = \frac{1}{\Omega} \int_{\Omega} f(\mathbf{x})\, \mathrm{d}\Omega \approx \frac{1}{\Omega} \sum_{i=1}^{N} f_i\, \Omega_i \quad , \tag{29}$$

where f_i and Ω_i are the function value and integration point volume, respectively, associated with the i-th integration point within a given volume Ω that contains a total of N integration points. Higher statistical moments of the fields, such as standard deviations, can be evaluated in analogy.

Applications of the FEM to micromechanical problems tend to fall into four main groups, compare figure 1. In most published works the phase arrangements are discretized by "standard" continuum elements, the mesh being designed in such a way that element boundaries coincide with all interfaces between constituents. Such an approach has the advantage that in principle any microgeometry can be handled, that interfaces can be modeled by interface elements if required, and that readily available commercial FE packages may be used. Meshing is often a considerable hurdle in using this discretization strategy, with complex phase configurations requiring sophisticated preprocessors. When volume elements of regular shape, e.g., right hexahedra, are used, intersections between the faces of the cell and phase boundaries at very acute angles can be especially difficult to handle, with suboptimal element shapes leading to unfavorably conditioned stiffness matrices. The capability of providing for mesh refinements where they are required is a major strength of standard FE discretizations of inhomogeneous solids, but tends to lead to numerical models of substantial size. Similar or identical meshes can be used for thermomechanical and thermal analysis. In studying cellular materials, shell and beam elements can often be employed to advantage.

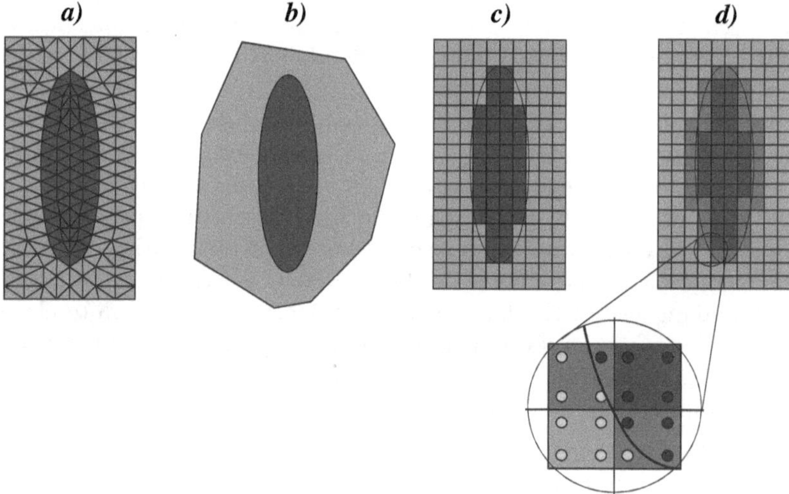

Figure 1. Sketch of FEM-based modeling strategies used in continuum micromechanics: a) discretization by standard elements, b) special hybrid elements, c) pixel/voxel discretization, d) "multiphase elements" (Böhm, 2004b).

Alternatively, a smaller number of special hybrid elements may be used, which are specifically formulated to model the displacement, stress and strain fields in a region consisting of a single inhomogeneity together with the surrounding matrix on the basis of some appropriate analytical theory.

This modeling strategy is exemplified by the Voronoi Finite Element Method of Ghosh et al. (1996), in which the mesh for the hybrid elements is obtained by Voronoi tesselations based on the positions of the reinforcements. Large planar multi-inhomogeneity arrangements can be analyzed this way using a limited number of (albeit rather complex) elements, and good accuracy as well as significant gains in efficiency have been claimed.

A third discretization strategy has been employed especially in cases where the phase arrangements to be studied are based on experimentally obtained digital representations of actual microgeometries. It uses a topologically and geometrically regular mesh that consists of rectangular or hexahedral elements of fixed size and has the same resolution as the digital data. Each element is assigned to one of the constituents by operations such as thresholding of the grey values of the corresponding pixel or voxel, respectively. Such models have the advantage of being highly suitable for (semi-) automatic generation from appropriate experimental data and of avoiding possible ambiguities in detecting and smoothing interfaces in the digital data. Obviously "voxel element" or "digital image based" strategies lead to ragged phase boundaries, which have, however, been found acceptable in many situations (Guldberg et al., 1998).

Regular FE meshes are also used in a fourth approach, where phase properties are assigned to Finite Elements at the integration point level ("multiphase elements"). Essentially, this amounts to trading off ragged boundaries at element edges for smeared-out (and typically degraded) microfields within those elements that contain a phase boundary, because stress or strain discontinuities within elements cannot be adequately handled by standard FE shape functions. With respect to the element stiffnesses the latter concern can be much reduced by overintegrating elements containing phase boundaries, which provides satisfactory approximations of integrals involving non-smooth displacements (Zohdi and Wriggers, 2001).

3.2 Periodic Homogenization

An important group of discrete microstructure approaches involves periodic model materials, the effective properties of which are used to approximate the behavior of actual, non-periodic inhomogeneous materials. This modeling strategy is referred to as periodic homogenization or as the periodic microfield approach.

Unit Cells Periodic homogenization involves studying unit cells, i.e., volume elements with periodic phase arrangements that tile space by translation and provide a complete description of periodic media of infinite extension. The cell volume of the smallest unit of periodicity is uniquely defined for a given periodic medium, whereas the translation vectors $\mathbf{p}_l$ describing such volumes and the shapes of unit cells of minimal size ("minimum unit cells") are non-unique. Computer generated periodic volume elements are often set up such that the translation vectors are orthogonal, which gives rise to unit cells that are rectangles or right hexahedra, compare fig. 2, and to some extent facilitates the application of periodicity boundary conditions. The faces of such cells typically intersect phase boundaries.

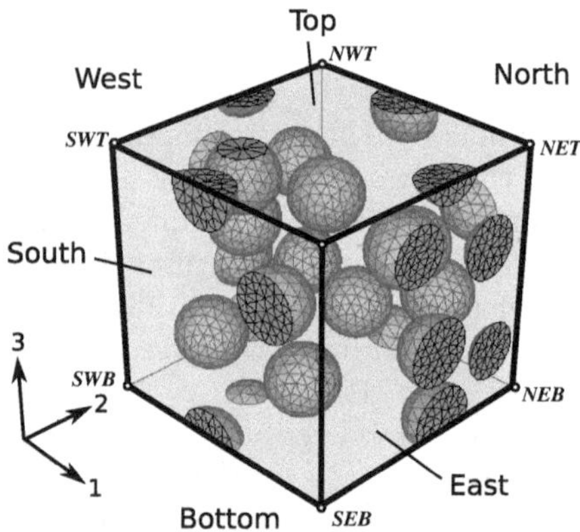

Figure 2. Cube-shaped periodic unit cell containing 15 randomly positioned spherical particles of equal size at a volume fraction of $\xi^{(i)}$=0.15. Designators of the six faces (East, West, North, South, Top, Bottom) and of the vertices are given (Pahr and Böhm, 2008).

In general, however, the periodicity translation vectors $\mathbf{p}_l$ are not orthogonal and, accordingly, unit cells of irregular shape may be chosen. Such unit cells can be used for avoiding local configurations that are difficult to mesh, e.g., at intersections between material interfaces and cell faces. Figure 3 shows a two-dimensional periodic array of circles of equal radius, which can be used to model a transverse cross section of some material reinforced by continuous aligned fibers. In it four different, but equivalent unit cells are marked, each of which contains four fibers. Whereas the faces of unit cells do not have to be planar, the condition must be fulfilled that the surface of any unit cell can be split into regions ${}^k\Gamma$, each of which consists of two parallel surface elements, k^- and k^+. The positions of the two surface elements making up such a pair must differ by a vector ${}^k\mathbf{s}$ that is a linear combination of the periodicity translation vectors, ${}^k\mathbf{s} = \sum_l {}^k c_l \mathbf{p}_l$, where the

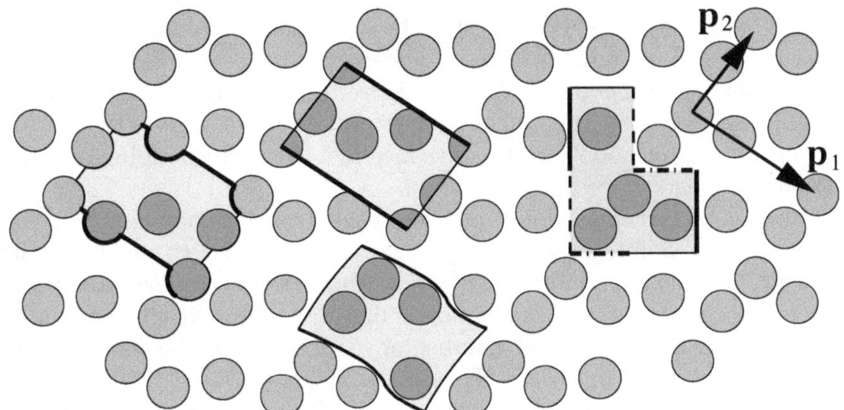

Figure 3. Different minimum unit cells for a two-dimensional periodic matrix–inclusion medium with two (slightly) non-orthogonal translation vectors $\mathbf{p}_1$ and $\mathbf{p}_2$. Paired face segments are marked by identical line styles and inhomogeneities inside the unit cells are set off by darker shading.

${}^k c_l$ are integer numbers. In fig. 2, for example, there are three such regions that coincide with pairs of cell faces, viz., East and West, North and South, as well as Bottom and Top.

Boundary Conditions When an appropriate unit cell has been chosen, it must be subjected to periodicity boundary conditions. This implies that the FE-discretizations on each pair of surface elements that make up a region ${}^k\Gamma$ must be compatible, i.e., that the same mesh is used on both of them. For such a pair of surface elements the boundary conditions for the mechanical problem can be expressed as

$$\Delta\,{}^k\mathbf{u} = {}^{k^+}\mathbf{u}(\mathbf{x}) - {}^{k^-}\mathbf{u}(\mathbf{x}) = \langle\varepsilon\rangle * {}^k\mathbf{s} \quad , \tag{30}$$

$\langle\varepsilon\rangle$ being the macroscopic strain, which is prescribed in displacement controlled analysis and which is to be determined in load controlled analysis. These conditions enforce a “seamless fit” between neighboring unit cells for all possible deformed states. The analogous relationship for the thermal conduction problem takes the form

$$\Delta\,{}^kT = {}^{k^+}T(\mathbf{x}) - {}^{k^-}T(\mathbf{x}) = \langle\mathbf{d}\rangle\;{}^k\mathbf{s} \quad , \tag{31}$$

where $\langle\mathbf{d}\rangle$ is the macroscopic temperature gradient. These boundary conditions may be interpreted as slaving the behavior of one face of each pair of surface elements ${}^{k}\Gamma$, called the slave face, to the other one, known as the master face. The increments of the macroscopic displacements and temperatures over the unit cell, $\langle \boldsymbol{\varepsilon} \rangle * {}^{k}\mathbf{s}$ and $\langle \mathbf{d} \rangle\, {}^{k}\mathbf{s}$, respectively, can be evaluated from the differences in the displacements and temperatures between pairs of nodes, a "master node" on the slave face and the corresponding reference node on the master face.

If the phase arrangement (and thus each translational unit cell) has planes, axes or points of symmetry, these can be used to define smaller unit cells that use symmetry and point symmetry boundary conditions, respectively. Mirror symmetry is not maintained through all deformation states, so that only a limited number of load cases can be handled with symmetry boundary conditions. For details see, e.g., Böhm (2004b, 2007).

In displacement and temperature based FE formulations, periodicity can be enforced in terms of these variables only, whereas periodicity in terms of stresses and fluxes is fulfilled approximately. All of the boundary conditions discussed above can be realized via linear multi-point constraints (i.e., linear equations linking a number of degrees of freedom), which are available in most commercial FE-packages. When doing load or flux controlled analysis, appropriate displacement or temperature degrees of freedom must be fixed to avoid rigid body modes.

Linking Microscale and Macroscale The most versatile and elegant strategy for linking the macroscale and the microscale in periodic microfield models is based on a mathematical framework known as asymptotic homogenization or homogenization theory, where the governing equations are explicitly formulated in terms of macroscopic and microscopic coordinates. Asymptotic homogenization directly provides the macroscopic elasticity or conductivity tensors of a unit cell, but typically requires special FE codes; for details see, e.g. Jansson (1992), Hassani and Hinton (1999), Chung et al. (2001) or Matt and Cruz (2008). It is also suitable for evaluating tangent modulus tensors in nonlinear analysis, compare, e.g., Ghosh et al. (1996), and for second-order homogenization problems explicitly involving stress and strain gradients (Kouznetsova et al., 2002).

Alternatively, the structure of the periodicity boundary conditions, eqns. (30) and (31), can be made use of, the macroscopic response being introduced via pairs of master and reference nodes. This approach was termed the "method of macroscopic degrees of freedom" by Michel et al. (1999). Displacement controlled analysis then takes the form of prescribing appropriate displacements to the master nodes, whereas for load controlled analysis concentrated forces must be applied to them. Following Smit et al. (1998), these concentrated forces can be obtained as the surface integrals of the homogeneous "applied traction" vector, $\boldsymbol{\sigma}^{\mathrm{a}} * \mathbf{n}_{\Gamma}(\mathbf{x})$, over the appropriate slave face, $\mathbf{n}_{\Gamma}(\mathbf{x})$ being the surface normal vector. For example, the West and East faces of the cell shown in fig. 2 form a master–slave pair with reference node SWB and master node SEB. The concentrated force acting on node SEB due to a prescribed macroscopic stress tensor $\boldsymbol{\sigma}^{\mathrm{a}}$, $\mathbf{f}_{\mathrm{SEB}}$, then evaluates as

$$\mathbf{f}_{\mathrm{SEB}} = \int_{\Gamma_{\mathrm{E}}} \boldsymbol{\sigma}^{\mathrm{a}} * \mathbf{n}_{\Gamma}(\mathbf{x}) \, \mathrm{d}\Gamma \quad . \tag{32}$$

The corresponding heat flow (or concentrated heat flux) in 1-direction, $q_{\mathrm{SEB},1}$ takes the form

$$q_{\mathrm{SEB},1} = \int_{\Gamma_{\mathrm{E}}} \mathbf{q}^{\mathrm{a}} \, \mathbf{n}_{\Gamma}(\mathbf{x}) \, \mathrm{d}\Gamma \quad , \tag{33}$$

where $\mathbf{q}^{\mathrm{a}}$ is the prescribed macroscopic heat flux. Within standard FE programs these concentrated loads can be easily prescribed. Outside the small strain regime eqn. (32) must, however, be evaluated for the current deformed configuration.

Averaged fields as well as phase averages of fields can be computed from unit cell models by using eqns. (1) and (29). In addition, macroscopic strains, stresses and gradients can be evaluated from the displacements, reaction forces and temperatures, respectively, of the master nodes.

The method of macroscopic degrees of freedom can be extended to mechanical analysis involving finite strains by approximating the homogenized deformation gradient ${}^{t}_{0}\boldsymbol{\phi}$ via the relations

$${}_{t}\mathbf{p}_{l} = {}^{t}_{0}\boldsymbol{\phi} * {}_{0}\mathbf{p}_{l} \quad , \tag{34}$$

where ${}_{0}\mathbf{p}_{l}$ and ${}_{t}\mathbf{p}_{l}$ are the l-th periodicity translation vectors at states 0 and t, respectively, see, e.g., Huber (2008).

Periodic homogenization is at present the most flexible and best understood discrete microstructure approach to continuum micromechanics. Its main limitations lie in the modeling of damage and failure, where diffuse damage can be handled but periodic fields are unrealistic once localization has set in, in dynamic problems, where periodic phase arrangements in general and unit cells specifically act as filters that can strongly restrict the passage of waves, and in stability problems, where similar effects can affect buckling modes. Aside from these issues periodic homogenization is capable of modeling the behavior of essentially any material for which a valid unit cell can be set up.

3.3 Windowing Methods

The aim of windowing methods consists in estimating or bounding the macroscopic properties of inhomogeneous materials on the basis of (non-periodic) samples of small size. The windows typically are volume elements of simple shape, are extracted at random positions and with random orientations from an inhomogeneous medium, and are not sufficiently big to be proper RVEs. Because the results of windowing pertain to individual samples rather than to a material, they are referred to as apparent (rather than effective) properties.

The evaluation of the macroscopic properties of a volume element can be based on the expression

$$\int_\Gamma \left[\mathbf{t}(\mathbf{x}) - \langle\boldsymbol{\sigma}\rangle * \mathbf{n}_\Gamma(\mathbf{x})\right]^\mathrm{T} \left[\mathbf{u}(\mathbf{x}) - \langle\boldsymbol{\varepsilon}\rangle * \mathbf{x}\right] \mathrm{d}\Gamma = 0 \quad , \tag{35}$$

which can be obtained from the Hill condition, eqn. (2), compare (Hazanov, 1998). Here Γ stands for the surface of the volume element and $\mathbf{t}(\mathbf{x}) = \boldsymbol{\sigma}(\mathbf{x}) * \mathbf{n}_\Gamma(\mathbf{x})$ is the surface traction vector. The analogous relationship for thermal conduction takes the form

$$\int_\Gamma \left[[\mathbf{q}(\mathbf{x}) - \langle\mathbf{q}\rangle]^\mathrm{T} \mathbf{n}_\Gamma(\mathbf{x})\right] \left[T(\mathbf{x}) - \langle\mathbf{d}\rangle^\mathrm{T} \mathbf{x}\right] \mathrm{d}\Gamma = 0 \quad , \tag{36}$$

(Jiang et al., 2002a). In general there are four ways of fulfilling these equations, three of them being based on uniform boundary conditions (Hazanov and Amieur, 1995; Ostoja-Starzewski, 2006).

First, the traction term in eqn. (35) or the normal flux term in eqn. (36) can be set equal to zero by specifying appropriate Neumann boundary conditions for the tractions $\mathbf{t}(\mathbf{x})$ or the normal fluxes $q_\mathrm{n}(\mathbf{x}) = \mathbf{q}^\mathrm{T}(\mathbf{x})\, \mathbf{n}_\Gamma(\mathbf{x})$. This can be enforced by prescribing a given macroscopically homogeneous stress tensor $\langle\boldsymbol{\sigma}\rangle^\mathrm{a}$ or flux vector $\langle\mathbf{q}\rangle^\mathrm{a}$ on all faces of the volume element,

$$\mathbf{t}(\mathbf{x}) = \langle\boldsymbol{\sigma}\rangle^\mathrm{a} * \mathbf{n}_\Gamma(\mathbf{x}) \qquad q_\mathrm{n}(\mathbf{x}) = \langle\mathbf{q}\rangle^{\mathrm{a}\mathrm{T}} \mathbf{n}_\Gamma(\mathbf{x}) \qquad \forall \mathbf{x} \in \Gamma \quad , \tag{37}$$

leading to statically uniform (SUBC) or uniform Neumann (UNBC) boundary conditions, respectively.

Second, the displacement or temperature terms in eqns. (35) and (36) can be set to zero by imposing a given macroscopically homogeneous strain tensor $\langle\boldsymbol{\varepsilon}\rangle^\mathrm{a}$ or temperature gradient vector $\langle\mathbf{d}\rangle^\mathrm{a}$ on all boundaries,

$$\mathbf{u}(\mathbf{x}) = \langle\boldsymbol{\varepsilon}\rangle^\mathrm{a} * \mathbf{x} \qquad T(\mathbf{x}) = \langle\mathbf{d}\rangle^{\mathrm{a}\mathrm{T}} \mathbf{x} \qquad \forall \mathbf{x} \in \Gamma \quad . \tag{38}$$

This results in kinematically uniform (KUBC) or uniform Dirichlet (UDBC) boundary conditions.

Third, mixed uniform boundary conditions (MUBC) may be specified, which enforce the vector or scalar product under the integral to vanish separately for each face Γ_k of the surface

$$\begin{aligned} \left[\mathbf{t}(\mathbf{x}) - \langle\boldsymbol{\sigma}\rangle * \mathbf{n}_\Gamma(\mathbf{x})\right]^{\mathrm{T}}\left[\mathbf{u}(\mathbf{x}) - \langle\boldsymbol{\varepsilon}\rangle * \mathbf{x}\right] \mathrm{d}\Gamma = 0 \\ \left[q_{\mathrm{n}}(\mathbf{x}) - \langle\mathbf{q}\rangle^{\mathrm{T}}\mathbf{n}_\Gamma(\mathbf{x})\right]\left[T(\mathbf{x}) - \langle\mathbf{d}\rangle^{\mathrm{T}}\mathbf{x}\right] \mathrm{d}\Gamma = 0 \qquad \forall\mathbf{x} \in \Gamma_k\,. \end{aligned} \tag{39}$$

In the mechanical case this involves appropriate combinations of traction and strain components that are uniform over a given face of the volume element, but not macroscopically homogeneous. Mixed uniform boundary conditions that fulfill eqn. (35) must be orthogonal in the fluctuating contributions (Hazanov and Amieur, 1995).

Finally, the fluctuations of non-uniform boundary fields can be made to cancel out by pairing parallel faces of the volume element such that they show identical fluctuations but surface normals of opposite orientations. This strategy, which does not involve homogeneous fields, is implemented by the periodicity boundary conditions (PBC) used in periodic homogenization, see section 3.2.

Macrohomogeneous boundary conditions following eqns. (37) and (38) can be shown to give rise to lower and upper estimates, respectively, for the overall elastic stiffness and thermal conductivity of a given volume element (Nemat-Nasser and Hori, 1993). Ensemble averages of such estimates obtained from windows of comparable size provide lower and upper bounds on the overall apparent tensors of these volume elements. Hierarchies of bounds can be generated from sets of windows of different sizes (Ostoja-Starzewski, 1998; Huet, 1999), bringing out effects of the size of the volume elements.

Equations (39) can be fulfilled by a range of different MUBC, resulting in different estimates for the apparent macroscopic tensors. A specific set of mixed uniform boundary conditions that avoids prescribing nonzero boundary tractions was proposed by Pahr and Zysset (2008) for obtaining the apparent elastic tensors of cellular materials. Table 2 lists these six load cases in a formulation suitable for windows in the shape of right hexahedra aligned with the coordinate system that have edge lengths l_1, l_2 and l_3 in the 1-, 2- and 3-directions, respectively. The components of the prescribed strain tensor are denoted as $\varepsilon_{ij}^{\mathrm{a}}$ and those of the prescribed traction vector as t_i^{a}. When applied to periodic volume elements with phase arrangements that show orthotropic or higher symmetry, these MUBC were found to give the same predictions for the macroscopic elasticity tensor as periodic ho-

Table 2. The six linearly independent uniform strain load cases constituting the periodicity compatible mixed uniform boundary conditions (PMUBC) in elasticity (Pahr and Zysset, 2008). The nomenclature of the volume element faces follows fig. 2.

	tensile 1	tensile 2	tensile 3
East	$u_1 = \varepsilon^a_{11} l_1/2$ $t^a_2 = t^a_3 = 0$	$u_1 = 0$ $t^a_2 = t^a_3 = 0$	$u_1 = 0$ $t^a_2 = t^a_3 = 0$
West	$u_1 = -\varepsilon^a_{11} l_1/2$ $t^a_2 = t^a_3 = 0$	$u_1 = 0$ $t^a_2 = t^a_3 = 0$	$u_1 = 0$ $t^a_2 = t^a_3 = 0$
North	$u_2 = 0$ $t^a_1 = t^a_3 = 0$	$u_2 = \varepsilon^a_{22} l_2/2$ $t^a_1 = t^a_3 = 0$	$u_2 = 0$ $t^a_1 = t^a_3 = 0$
South	$u_2 = 0$ $t^a_1 = t^a_3 = 0$	$u_2 = -\varepsilon^a_{22} l_2/2$ $t^a_1 = t^a_3 = 0$	$u_2 = 0$ $t^a_1 = t^a_3 = 0$
Top	$u_3 = 0$ $t^a_1 = t^a_2 = 0$	$u_3 = 0$ $t^a_1 = t^a_2 = 0$	$u_3 = \varepsilon^a_{33} l_3/2$ $t^a_1 = t^a_2 = 0$
Bottom	$u_3 = 0$ $t^a_1 = t^a_2 = 0$	$u_3 = 0$ $t^a_1 = t^a_2 = 0$	$u_3 = -\varepsilon^a_{33} l_3/2$ $t^a_1 = t^a_2 = 0$
	shear 12	shear 13	shear 23
East	$u_2 = \varepsilon^a_{21} l_1/2$ $u_3 = 0, t^a_1 = 0$	$u_3 = \varepsilon^a_{31} l_1/2$ $t^a_2 = 0, t^a_1 = 0$	$u_1 = 0$ $t^a_2 = t^a_3 = 0$
West	$u_2 = -\varepsilon^a_{21} l_1/2$ $u_3 = 0, t^a_1 = 0$	$u_3 = -\varepsilon^a_{31} l_1/2$ $t^a_2 = 0, t^a_1 = 0$	$u_1 = 0$ $t^a_2 = t^a_3 = 0$
North	$u_1 = \varepsilon^a_{12} l_2/2$ $u_3 = 0, t^a_2 = 0$	$u_2 = 0$ $t^a_1 = t^a_3 = 0$	$u_3 = \varepsilon^a_{32} l_2/2$ $u_1 = 0, t^a_2 = 0$
South	$u_1 = -\varepsilon^a_{12} l_2/2$ $u_3 = 0, t^a_2 = 0$	$u_2 = 0$ $t^a_1 = t^a_3 = 0$	$u_3 = -\varepsilon^a_{32} l_2/2$ $u_1 = 0, t^a_2 = 0$
Top	$u_3 = 0$ $t^a_1 = t^a_2 = 0$	$u_1 = \varepsilon^a_{13} l_3/2$ $u_2 = 0, t^a_3 = 0$	$u_2 = \varepsilon^a_{23} l_3/2$ $u_1 = 0, t^a_3 = 0$
Bottom	$u_3 = 0$ $t^a_1 = t^a_2 = 0$	$u_1 = -\varepsilon^a_{13} l_3/2$ $u_2 = 0, t^a_3 = 0$	$u_2 = -\varepsilon^a_{23} l_3/2$ $u_1 = 0, t^a_3 = 0$

mogenization. Accordingly, they were named "periodicity compatible mixed uniform boundary conditions" (PMUBC).

The concept of periodicity compatible mixed uniform boundary conditions can be extended to thermoelasticity by adding a load case that constrains all displacements normal to the faces of the volume element, sets all in-plane tractions to zero, and applies a uniform temperature increment ΔT. This allows to evaluate the volume averaged specific thermal stress tensor

$\langle \boldsymbol{\vartheta} \rangle$, from which the apparent thermal expansion tensor can be obtained as $\langle \boldsymbol{\alpha} \rangle = -\mathbf{C}^* \langle \boldsymbol{\vartheta} \rangle$.

Boundary conditions that show an analogous behavior to the above PMUBC were reported by Jiang et al. (2002a) for thermal conduction in two-dimensional orthotropic periodic media. Table 3 lists the three thermal load cases required for extracting the apparent thermal conduction tensor in the three-dimensional case, which make use of prescribed thermal gradient and flux components d_i^{a} and q_i^{a}, respectively.

Table 3. The three linearly independent uniform gradient load cases constituting the periodicity compatible mixed uniform boundary conditions in thermal conduction. The nomenclature of the faces follows fig. 2.

	thermal 1	thermal 2	thermal 3
East	$T = d_1^{\mathrm{a}} l_1/2$	$q_1^{\mathrm{a}} = 0$	$q_1^{\mathrm{a}} = 0$
West	$T = -d_1^{\mathrm{a}} l_1/2$	$q_1^{\mathrm{a}} = 0$	$q_1^{\mathrm{a}} = 0$
North	$q_2^{\mathrm{a}} = 0$	$T = d_2^{\mathrm{a}} l_2/2$	$q_2^{\mathrm{a}} = 0$
South	$q_2^{\mathrm{a}} = 0$	$T = -d_2^{\mathrm{a}} l_2/2$	$q_2^{\mathrm{a}} = 0$
Top	$q_3^{\mathrm{a}} = 0$	$q_3^{\mathrm{a}} = 0$	$T = d_3^{\mathrm{a}} l_3/2$
Bottom	$q_3^{\mathrm{a}} = 0$	$q_3^{\mathrm{a}} = 0$	$T = -d_3^{\mathrm{a}} l_3/2$

The PMUBC listed in tables 2 and 3 offer an attractive option for evaluating estimates of the macroscopic elasticity and conductivity tensors of periodic and non-periodic volume elements. The question if and to what extent they are applicable to obtaining approximations of the macroscopic behavior of microgeometries that show sub-orthotropic symmetry is explored in section 4.3.

The generation of lower and upper estimates by windowing using SUBC and KUBC can be shown to be valid in the context of nonlinear elasticity and deformation plasticity (Jiang et al., 2001, 2002b). PMUBC can be used for materials characterization of elastoplastic inhomogeneous materials (Pahr and Böhm, 2008), but lack the flexibility for handling general load paths in plasticity. The principal strength of windowing methods lies in providing an approach to studying the linear behavior of non-periodic volume elements.

3.4 Embedding Methods

Embedding methods combine a geometrically well resolved core region, in which the local fields can be evaluated at high detail, with an outer region that describes the inhomogeneous material in a simplified way and serves mainly for introducing loads into the core. The material behavior in the outer region may either be prescribed, e.g., via a suitable mean field model, or be evaluated self-consistently from the responses of the core, the latter approach being difficult to use properly for nonlinear studies. Embedding models cannot avoid boundary layers in both core and embedding region, where local perturbations of the fields are present. These boundary layers must be excluded from the evaluation of macroscopic quantities and volume averaged fields.

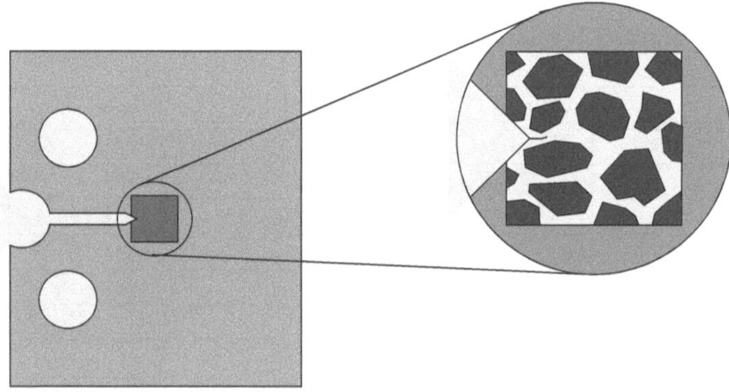

Figure 4. Schematic sketch of an embedded cell analysis of a crack in a tensile specimen of a composite material

The main strength of embedded cell models lies in their capability of describing non-periodic local phenomena in inhomogeneous materials, most importantly the regions around the tips of macrocracks as sketched in fig. 4. When combined with appropriate material models that can handle local damage and failure at the constituent level such models allow to simulate the progress of macrocracks in inhomogeneous materials, see, e.g., van der Giessen and Tvergaard (1994) or Ayyar and Chawla (2007). Other problems to which embedded cells can be applied to advantage include macroscopic interfaces involving at least one inhomogeneous material (Chimani et al., 1997) as well as other situations where gradients of loads or geometrical parameters are too high for standard homogenization methods.

4 Some Issues and Applications

4.1 Incremental Mori–Tanaka Methods for Elastoplastic Composites at Finite Strains

As noted in section 2.4, algorithmic enhancements have led to marked improvements in Incremental Mori–Tanaka (IMT) schemes, especially for macroscopically isotropic composites. Such methods can be extended to make them suitable for handling finite macroscopic strains, e.g., for forming simulations involving particle reinforced metal matrix composites (MMCs). In these materials elastic strains are small compared to the inelastic ones because the Young's modulus of the matrix typically is orders of magnitude larger than the yield stress. This allows to use the additive decomposition of strain rates as well as the choice of Cauchy stress and logarithmic strain as conjugate measures. Matrix plasticity is treated as the exclusive origin of the rotation of the material base reference system, the rotation of the matrix phase being assumed to be fully conveyed to the embedded elastic particles. Enhanced IMT schemes of this type can be implemented into Finite Element programs as micromechanically based constitutive models at the integration point level (Huber et al., 2007; Huber, 2008). In addition to enabling multi-scale forming simulations of components made of MMCs, such models allow comparing the predictions of IMT and unit cell approaches for non-radial load paths.

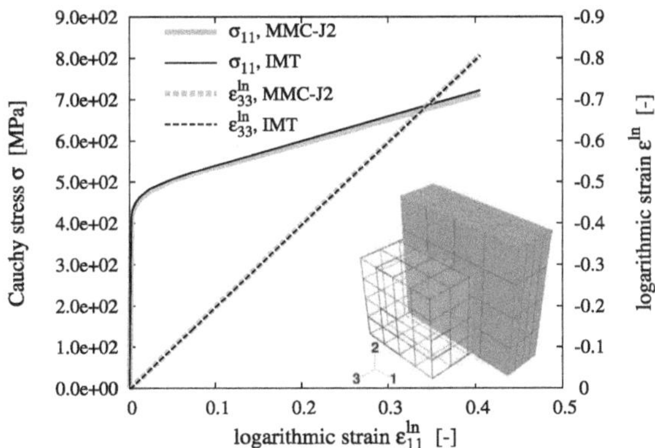

Figure 5. Comparison between predictions of a unit cell (MMC–J2) and a finite strain IMT model for the equi-biaxial tensile loading of a particle reinforced MMC (Huber, 2008). The macroscopic stress–strain diagram in the plane of loading and the normal strain are shown.

Figure 5 shows the macroscopic stress–strain responses under equi-biaxial tension predicted, on the one hand, with a periodic unit cell of the type depicted in fig. 2 and, on the other hand, with a single Finite Element employing an IMT model modified for finite strains. Both models pertain to an MMC consisting of elastic spherical particles with $E^{(\mathrm{i})} = 400$ GPa, $\nu^{(\mathrm{i})} = 0.19$ and $\xi^{(\mathrm{i})} = 0.2$ embedded in an elastoplastic matrix with elastic parameters $E^{(\mathrm{m})} = 100$ GPa and $\nu^{(\mathrm{m})} = 0.30$, J_2-plasticity and isotropic hardening. The macroscopic stress–strain diagram (σ_{11} vs. ε_{11}) as well as the evolution of the normal strain ε_{33} are shown, excellent agreement between the predictions of the two models being evident.

An axisymmetric FE model employing the IMT was used to simulate a Gleeble-type experiment, i.e., a compression test of a cylindrical specimen between two anvils, of the same material (Huber et al., 2007). The deformation gradients obtained at selected points were treated as mesoscopic responses and applied to a periodic unit cell. Good agreement between the IMT and periodic homogenization models was found in terms of the evolution of the stress components at these points. The computationally expensive unit cell approach also allows to extract the distributions of microfields in the constituents at given mesoscopic points. Figure 6 shows such a "spectrum" of the accumulated equivalent plastic strain in the ma-

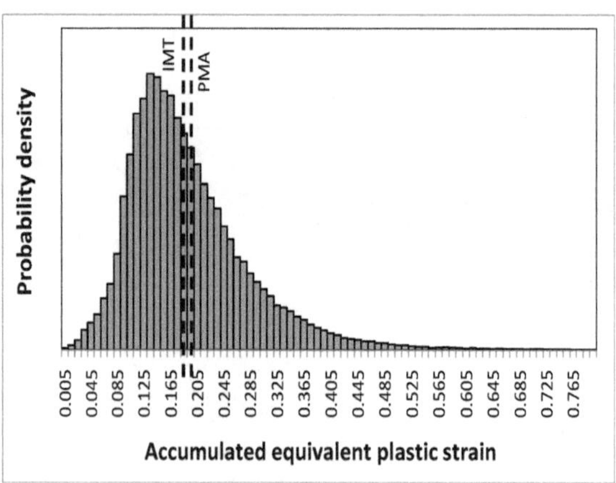

Figure 6. Unit cell prediction for the distribution of the accumulated equivalent plastic strain in the matrix at the center of an initially cylindrical MMC sample under uniaxial compression. Phase averages predicted by the unit cell model (PMA) and the finite strain IMT model are shown (Huber et al., 2007).

trix. These results pertain to the center of the sample and a macroscopic deformation gradient of $\phi_{22} = 0.88$. The mean value of this distribution can be seen to be slightly higher than the corresponding IMT mean field prediction.

Underestimating the equivalent stresses and strains is a typical behavior for incremental mean field methods that evaluate the equivalent stress from the phase averaged stress components and thus do not account for correlations between fluctuating contributions. Solutions to this problem were proposed, e.g., by Buryachenko (1996) and by Ju and Sun (2001).

As is evident from figure 7 stress fluctuations can also markedly influence the predicted phase averages of the maximum principal stress in the particles. This behavior can be especially marked at positions of high macroscopic symmetry such as the center of the sample in the present case, for which the IMT results do not hint at any propensity for brittle failure of the particles. The unit cell model, in contrast, picks up positive averages and long tails in the tensile range of the distribution of the local maximum principal stresses.

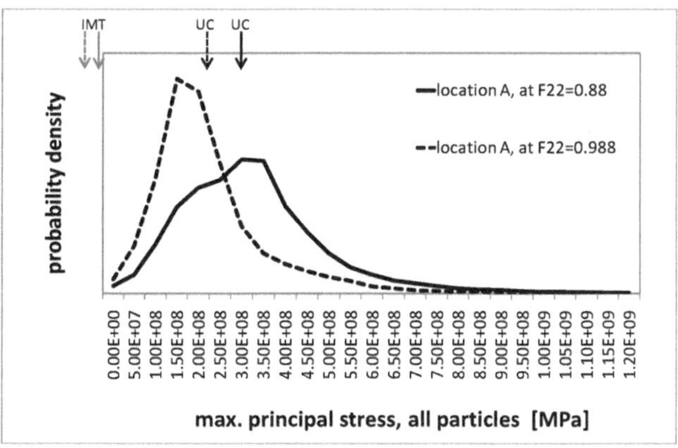

Figure 7. Unit cell prediction for the distribution of the maximum principal stress in the particles at the center of an initially cylindrical MMC sample under uniaxial compression. Phase averages predicted by the unit cell model (PMA) and the finite strain IMT model are shown (Huber et al., 2007).

4.2 Diamond Particle Reinforced Composites

Thermomechanical Behavior Aluminum or copper matrices reinforced with diamond particles are of considerable practical interest for use as heat sink materials because they promise, on the one hand, overall conductivities that exceed the corresponding properties of the matrix and, on the other hand, the possibility of tailoring the effective coefficients of thermal expansion (CTE) for a good match with active elements made of semiconductors. The marked thermal expansion mismatch between diamond ($\alpha^{(i)} \approx 1.5\times10^{-6}$ K^{-1} at room temperature) and matrix ($\alpha^{(m)} \approx 21.7\times10^{-6}$ K^{-1} for aluminum at room temperature), however, can lead to high microstresses in the constituents under thermal loading.

Periodic unit cells for describing diamond reinforced composites of particle volume fractions up to, say, 35% can be generated on the basis of arrangements of randomly positioned, non-interpenetrating spheres by inscribing a randomly oriented regular tetrakaidekahedron into each sphere. Figure 8 shows three unit cells of nominal particle volume fraction $\xi^{(i)} = 0.34$ that share the same particle center positions, the tetrakaidekahedra in the left and center cells differing in their orientation only (Nogales and Böhm, 2008). Alternatively, the replacement tensor algorithm discussed in section 2.7 can be used to address particle shape effects within the context of mean field approaches.

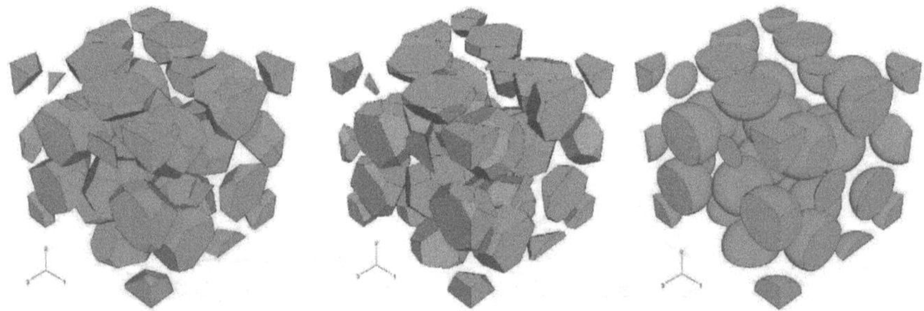

Figure 8. Three unit cells of nominal particle volume fraction $\xi^{(i)} = 0.34$ used for modeling the thermomechanical and thermal conduction behavior of diamond reinforced aluminum (Nogales and Böhm, 2008). The cells are referred to as UCDA (left), UCDB (center) and UCS (right).

Table 4 compares predictions for the effective Young's moduli, E^*, shear moduli, G^*, and bulk moduli, K^*, obtained with different models for the elastic behavior of diamond reinforced composites. The standard Mori–Tanaka (MTM,sph) and Torquato (1998b) three-point estimates (3PE) as well as periodic homogenization with unit cell UCS, compare fig. 8 (right), pertain to spherical particles. Regular tetrakaidekahedral particles were used with the MTM/RT model, the results of which were isotropized by

Table 4. Predictions for the effective elastic moduli [GPa] of diamond reinforced aluminum obtained for spherical (MTM,sph; 3PE,sph; UCS) and tetrakaidekahedral (MTM/RT; UCDA, UCDB) particles (Nogales, 2008).

model	E^*	G^*	K^*
MTM,sph	131.2	50.7	106.1
3PE,sph	137.2	53.3	107.3
UCS	135.7	52.7	106.6
MTM/RT	134.0	51.9	106.7
UCDA	141.7	55.3	108.0
UCDB	142.0	55.4	108.2

the Hershey–Kröner–Eshelby method (Ledbetter, 1984) because the "raw" MTM results show cubic elastic symmetry due to the shapes of the particles. Finally, results obtained with the multi-particle unit cells UCDA and UCDB are given. The diamond particles were treated as elastically isotropic and the elastic parameters for the constituents were chosen as $E^{(\mathrm{i})} = 1050$ GPa, $G^{(\mathrm{i})} = 477$ GPa, $E^{(\mathrm{m})} = 70$ GPa, and $G^{(\mathrm{m})} = 26$ GPa. Good agreement between the six sets of predicted moduli can be observed, the Mori–Tanaka estimates giving the lowest stiffnesses for both particle shapes. The polyhedral particles are predicted to give rise to somewhat higher macroscopic stiffnesses than the spherical ones by both the mean field and the discrete microfield methods. The differences between unit cells UCDA and UCDB are very small, i.e., the particle orientations have little influence on the overall elastic behavior.

Figure 9 (left) presents the predicted distribution of the accumulated equivalent plastic strain in the matrix after cooling down from a stress-free temperature of 450 K to room temperature. Most of the matrix can be seen to have yielded plastically in the cooled-down state, the distribution of the plastic strains in the matrix being very inhomogeneous. In fig. 9 (right) an analogous plot is given for a state obtained by subsequent heating by 100 K. Here the marked CTE mismatch and the low yield stress of the matrix have led to further plastic yielding, the values of the equivalent plastic strain being noticeably higher throughout the matrix.

Information on the evolution of the plastic strains of the matrix within the volume element can be obtained by evaluating the phase averages of the accumulated plastic strain and the corresponding standard deviations via eqn. (29). The resulting data can be plotted against the temperature as shown in fig. 10, which pertains to the heating-up process and uses error bars to represent the standard deviations of the equivalent plastic strain. For

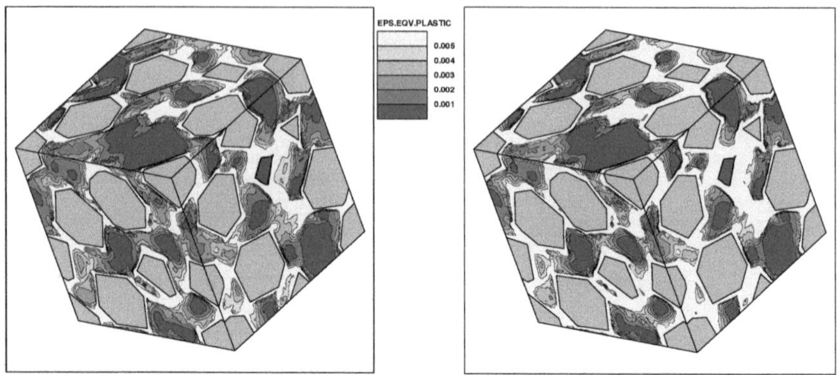

Figure 9. Accumulated equivalent plastic strains in the matrix of a diamond–aluminum composite subjected to cooling down from a stress-free temperature of 450 K to room temperature (left) and subsequent heating by 100 K (right) as predicted with unit cell UCDA, compare fig. 8 (left), with a particle volume fraction of $\xi^{(\mathrm{i})} = 0.34$ (Nogales, 2008).

temperature increases up to, say, 25°C the accumulated equivalent plastic strain remains constant, i.e., the matrix behaves thermoelastically. Further heating, however, leads to monotonous growth of both the phase average and

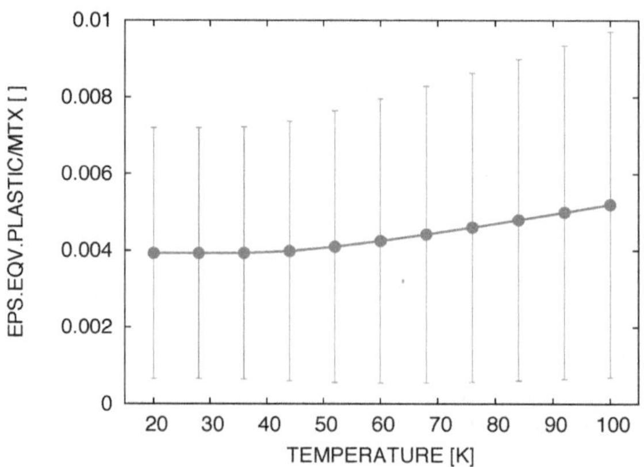

Figure 10. Evolution of phase average and standard deviation of the equivalent plastic strain in the matrix of an diamond–aluminum composite ($\xi^{(\mathrm{i})} = 0.34$) during heating from a cooled-down state as predicted with the multi-particle unit cell shown in fig. 9 (Nogales, 2008).

the standard deviation of the equivalent plastic strain, indicating progressive plastic yielding and growing intra-phase fluctuations of the plastic strain in the matrix.

Thermal Conduction In studying the thermal conduction behavior of diamond reinforced composites finite interfacial conductances at the interface between particles and matrix must be explicitly accounted for. Beyond this, the models and modeling techniques used for describing the thermomechanical responses of these composites are directly applicable.

Experimental studies (Ruch et al., 2006) indicate that the interfacial conductances of diamond particles in aluminum are inhomogeneous, values of $h_{\{100\}}$ = 100 MW/Km^2 and $h_{\{111\}}$ = 20 MW/Km^2 being assumed for the diamonds' {100} and {111} faces, respectively, in the following. For the given phase conductivities, $k^{(i)}$ = 1800 W/Km and $k^{(m)}$ = 237 W/Km, an "equivalent" homogeneous interfacial conductance of $h_{\mathrm{hom},200} \approx 27.7$ MW/Km^2 can be evaluated for dilute regular tetrakaidekahedra of 200 μm diameter (Nogales, 2008).

Figure 11 compares predictions of a number of models for the size dependence of the macroscopic conductivities of diamond–aluminum composites of particle volume fraction $\xi^{(i)} = 0.34$. Mori–Tanaka results for perfectly bonded and for fully debonded particles, the latter behaving like voids, are marked as "MTM, perfect" and "MTM, voids", respectively, and do not show a size effect. Predictions obtained by combining the replacement tensor algorithm with the Mori–Tanaka method, eqns. (27), are represented by the curves "MTM/RT, inh" and "MTM/RT, hom", which pertain to the use of inhomogeneous and equivalent homogeneous interfacial conductances, respectively. As is typical for Mori–Tanaka estimates, they do not consistently fall within the pertinent three-point bounds (Torquato and Rintoul, 1995), designated as "3PB, hom". In addition, numerical predictions based on unit cell UCDA, compare fig. 8, are shown for a number of particle diameters. At the critical particle diameter, $d_{\mathrm{cr}} \approx 19.7$ μm, the composite behaves like the matrix material and the upper and lower three-point bounds as well as the Mori-Tanaka predictions for homogeneous interfaces coincide. For the case considered here, the differences in the macroscopic conduction behavior due to particle shape and interfacial inhomogeneity effects can be seen to play a minor role only. For further details see Nogales (2008) as well as Nogales and Böhm (2008).

The diamond–aluminum system can also be studied by windowing models using the sets of boundary conditions discussed in section 3.3, non-periodic volume elements being available for higher particle volume fractions than were periodic unit cells (Flaquer et al., 2007). Nogales (2008) reported

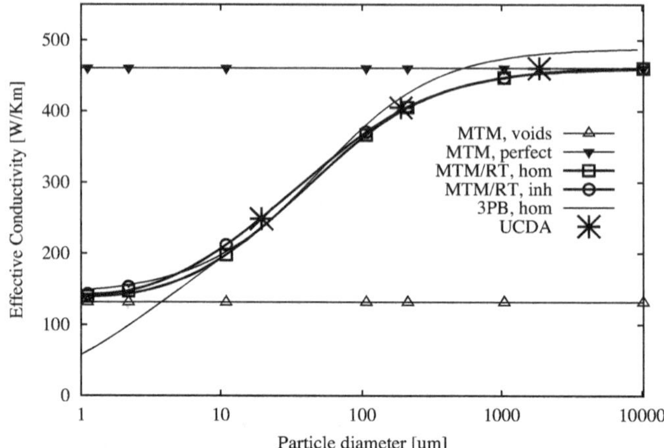

Figure 11. Predictions of the effective conductivity of diamond reinforced aluminum ($\xi^{(i)} = 0.34$). The Mori–Tanaka results for fully debonded (MTM, voids) and perfectly bonded (MTM, perfect) particles and the three-point bounds (3PB, hom) pertain to spherical inhomogeneities, whereas the MTM/RT predictions for particles with homogeneous and inhomogeneous interfaces and the unit cell results (UCDA) pertain to tetrakaidekahedral particles.

good agreement between the predictions obtained with the volume element shown in fig. 12 and effective conductivities obtained with the MTM/RT. Results obtained for unit cell UCDA with both periodic homogenization and windowing models are listed in table 6 and discussed in section 4.3.

In many real composites the reinforcements are not of exactly equal size, but follow some particle size distribution. When this size distribution can be cast into the form of a histogram-like description that provides volume fractions $\xi^{(i)}$ in terms of particle diameters and when the particle shapes can be approximated by spheres, eqns. (27) can be used to evaluate the (diagonal) components of the replacement tensors as functions of the pertinent particle diameters. $A^{(i,r)}$ and $K^{(i,r)}$, in turn, can be combined with eqns. (5), 6), (11) and (12) to obtain the Mori–Tanaka estimate

$$K^* = K^{(m)} + \frac{\sum\limits_{(i)\neq(m)} \xi^{(i)} [K^{(i,r)} - K^{(m)}] A^{(i,r)}_{\text{dil}}}{\xi^{(m)} + \sum\limits_{(i)\neq(m)} \xi^{(i)} A^{(i,r)}_{\text{dil}}} \tag{40}$$

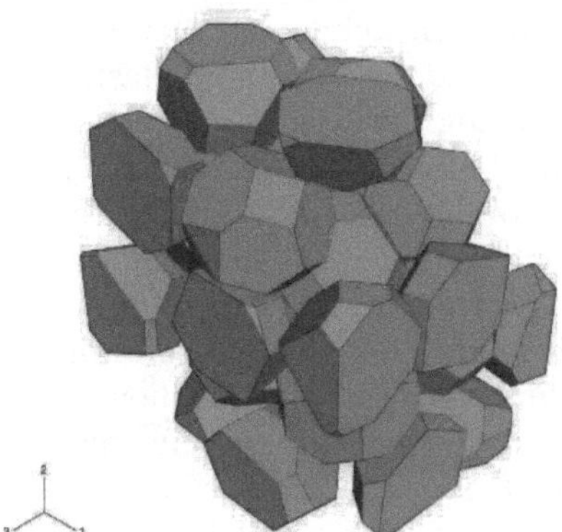

Figure 12. Non-periodic volume element for a diamond–aluminum composite of particle volume fraction $\xi^{(i)} = 0.445$ (Nogales, 2008).

for the isotropic macroscopic conductivity of a composite reinforced by equiaxed particles with finite interfacial conductances (Böhm and Nogales, 2008). Obviously, this model treats particles of different diameters as different inclusion phases. The resulting simple algebraic relationship provides a handy method for approximating the size effect in conduction due to finite interfacial conductances, e.g., for the practically relevant case of bimodal size distributions of equiaxed particles.

4.3 Windowing Methods Using PMUBC

As discussed in section 3.3 windowing methods using periodicity compatible mixed uniform boundary conditions (PMUBC) were reported to give identical results to periodic homogenization for periodic microstructures of orthotropic symmetry in elasticity (Pahr and Zysset, 2008) and thermal conduction (Jiang et al., 2002a). However, actual volume elements typically are too small to be proper RVEs and, as a consequence, tend to show some sub-orthotropic contributions to their overall symmetry even if they aim at modeling isotropic materials. The question whether and to what extent PMUBC are suitable for studying such microgeometries can be assessed by comparing the responses of periodic phase arrangements that are subjected, on the one hand, to PMUBC and, on the other hand, to periodicity boundary conditions (PBC). The latter provide well understood reference estimates that are not subject to restrictions in terms of symmetry.

This testing strategy was first applied to a multi-particle unit cell containing 15 randomly positioned spherical inhomogeneities at a volume fraction of $\xi^{(\mathrm{i})} = 0.15$, see fig. 2. This volume element had been used by Böhm et al. (2004) to study particle reinforced metal matrix composites. Both PBC and PMUBC were used to obtain full elasticity tensors, symmetrization of the raw results being necessary in the latter case. Approximations to the orthotropic contributions to these elasticity tensors were generated by setting to zero the non-orthotropic coupling terms, which are about an order of magnitude smaller than the orthotropic terms. This procedure was carried out for elastic contrasts, $E^{(\mathrm{i})}/E^{(\mathrm{m})}$, ranging between 2 and 30, and the macroscopic elastic moduli were evaluated from the "orthotropized" tensors, see Pahr and Böhm (2008). Figure 13 shows the relative differences of the predicted Young's, shear and bulk moduli with respect to the reference solutions obtained with PBC. The upper estimates obtained with kinematically uniform boundary conditions (KUBC) and the three-point bounds (Torquato, 2002) evaluated with statistical parameters pertaining to hard spheres of equal size (Torquato et al., 1987; Miller and Torquato, 1990) are also shown.

In fig. 13 direction averaged predictions for the Young's, shear and bulk moduli are marked by symbols and the levels of anisotropy are given by error bars where applicable. The results obtained with PMUBC differ by no more than 1.7% from the reference values in the case of the Young's moduli and by much less for the shear and bulk moduli. The PMUBC can be seen to lead to slightly higher deviations from macroscopic elastic isotropy than do the PBC. Both sets of estimates are compatible with the three-point bounds evaluated for monodisperse non-interpenetrating spheres, and the PBC data nearly coincide with the three-point lower bounds. The upper estimates obtained with KUBC, in contrast, differ markedly from the reference data,

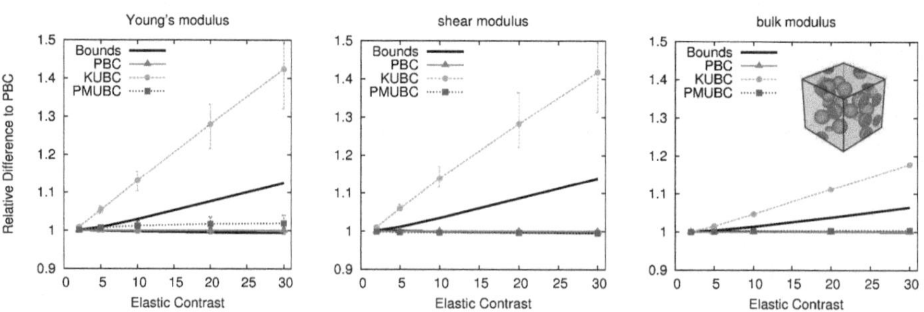

Figure 13. Relative differences in the macroscopic elastic responses of the volume element depicted in fig. 2 predicted numerically by PBC, KUBC and PMUBC and analytically by three-point bounds (Pahr and Böhm, 2008).

fall clearly outside the bounds, and show a marked anisotropy in the Young's and shear moduli. For windowing using KUBC the deviations from the PBC predictions are strongly correlated with the elastic contrast $E^{(i)}/E^{(m)}$, whereas for the estimates obtained with PMUBC they are nearly constant for elastic contrasts beyond ten.

By using the isotropic term of a generalized spherical harmonics expansion of the orthotropic elongation and bulk modulus orientation distribution functions developed by He and Curnier (1997), the isotropic elasticity tensors can be found that are closest to the orthotropized ones. For the unit cell shown in fig. 2 the maximum difference between the components of these macroscopic elastic tensors obtained with PBC and PMUBC is some 1.2% (Pahr and Böhm, 2008).

In a second step four groups of six periodic phase arrangements were generated, two of which pertain to particle volume fractions of $\xi^{(i)} \approx 0.15$ and two to $\xi^{(i)} \approx 0.40$, respectively. Unit cells containing 15 and 30 particles were used for each volume fraction. Each group comprised 6 microgeometries consisting of randomly positioned spheres of equal size (Gándara, 2008; Böhm et al., 2008). Table 5 lists the ensemble averages of the bulk and shear moduli obtained with PBC and PMUBC, respectively, together with the three-point bounds for impenetrable spheres of equal size. Again, very close agreement between periodic homogenization and windowing using PMUBC is evident, although some results for the shear modulus obtained with the latter approach are slightly below the three-point bounds.

Table 5. Ensemble averages of the shear and bulk moduli [GPa] obtained with periodic homogenization (PBC) and PMUBC from four groups of six unit cells each (Gándara, 2008). The pertinent three-point bounds are listed for comparison.

Shear Moduli	3PB	PBC	PMUBC
ξ=0.147, 15 particles	33.99 / 34.55	34.01	33.94
ξ=0.145, 30 particles	33.88 / 34.43	33.94	33.87
ξ=0.397, 15 particles	51.68 / 55.29	52.82	52.57
ξ=0.395, 30 particles	51.50 / 55.07	52.25	52.06
Bulk Moduli	3PB	PBC	PMUBC
ξ=0.147, 15 particles	68.21 / 68.56	68.17	68.27
ξ=0.145, 30 particles	68.06 / 68.40	68.06	68.14
ξ=0.397, 15 particles	91.00 / 93.32	91.33	91.73
ξ=0.395, 30 particles	90.78 / 93.08	90.95	91.29

For thermal conduction analysis comparisons between periodic homogenization and windowing using the PMUBC listed in table 3 were carried out on the basis of the unit cell shown in fig. 8 (left), which contains 20 randomly positioned and oriented regular tetrakaidekahedral particles of equal size (Nogales, 2008). Inhomogeneous interfacial conductances as discussed in section 4.2 were prescribed at the interfaces between matrix and reinforcements. Table 6 shows the predictions obtained with UNBC, PMUBC, PBC, and UDBC for this periodic volume element together with results from the Mori–Tanaka method employing the replacement tensor algorithm discussed in section 2.7. Because the finite interfacial conductances give rise to a particle size effect, four different particle sizes were studied. In each case the UNBC lead to lower and the UDBC to upper estimates, the relative difference being largest for the smallest particles, which approach the behavior of holes and thus lead to a high conductivity contrast. The predictions obtained with PMUBC and PBC differ by less than 3%.

Table 6. Comparison of effective and apparent conductivities [W/Km] obtained by periodic homogenization (PBC), windowing using uniform Dirichlet (UDBC), uniform Neumann (UNBC), and periodicity compatible mixed uniform (PMUBC) boundary conditions as well as the MTM/RT for four different particle sizes (Nogales, 2008).

	UNBC	PMUBC	PBC	UDBC	MTM/RT
$r = 10^{-6}$m	130.4	151.3	149.6	287.8	151.3
$r = 10^{-5}$m	246.5	256.6	249.2	364.3	249.2
$r = 10^{-4}$m	390.7	411.9	405.1	484.1	401.0
$r = 10^{-3}$m	441.0	466.1	460.4	529.2	453.6

These results show that PMUBC are well suited to evaluating the overall elastic and thermal conduction behavior of periodic unit cells that display considerable sub-orthotropic contributions to their overall symmetry. This property of PMUBC also holds for volume elements that are clearly too small to be RVEs, as is the case in table 6, where the lower and upper estimates obtained with UNBC and UDBC, respectively, differ by 20% to 120%. On the basis of the results presented in this section it can also be expected that PMUBC are valid for studying the macroscopic elastic behavior of non-periodic volume elements with overall symmetries that deviate to some extent from orthotropy.

4.4 Porous and Cellular Materials

For many purposes porous and cellular materials can be treated as inhomogeneous materials one constituent of which has vanishing stiffness and conductivity. This approximation is not valid if the void spaces are saturated with a liquid, a situation that is studied in poromechanics, see, e.g., Dormieux and Ulm (2005), or when the pressure of a gas filling the voids influences the mechanical response. In heat transfer analysis gas in the voids may give rise to conduction or convection and at high temperatures radiative heat transport may occur. Such effects are not considered in the following.

Porous materials Here porous materials are understood to have void volume fractions considerably smaller than 50%, and they typically show matrix–inclusion topology. For small strain elasticity they can be modeled by standard mean field theories by letting the elasticity and conductivity tensors of the inhomogeneities vanish. For the Mori–Tanaka method this leads to expressions of the type

$$\begin{aligned} \mathbf{E}^* &= (1-\xi^{(\mathrm{i})})\mathbf{E}^{(\mathrm{m})}[\mathbf{I}-\mathbf{S}^{(\mathrm{i,m})}][\mathbf{I}-\xi^{(\mathrm{m})}\mathbf{S}^{(\mathrm{i,m})}]^{-1} \\ \mathcal{K}^* &= (1-\xi^{(\mathrm{i})})\mathcal{K}^{(\mathrm{m})}[\mathcal{I}-\mathcal{S}^{(\mathrm{i,m})}][\mathcal{I}-\xi^{(\mathrm{m})}\mathcal{S}^{(\mathrm{i,m})}]^{-1} \quad , \end{aligned} \tag{41}$$

which correspond to Hashin–Shtrikman upper bounds. Classical self-consistent estimates for macroscopically isotropic materials break down for void volume fractions exceeding approximately one third, where the voids start to percolate through the microstructure and cohesion of the material is lost.

Elastoplastic porous materials have been the subject of a considerable number of micromechanical studies due to their relevance to the ductile damage and failure behavior of metals. Micromechanically based models have been employed for describing the growth of pre-existing voids, compare (Gurson, 1977; Gologanu et al., 1997; Kailasam et al., 2000). Generally, the underlying modeling concepts are closely related to the homogenization for particle reinforced composites, the main difference being that at finite macroscopic strains the shapes of the voids may evolve significantly through the loading history. Studies using axisymmetric cell models showed that for tensile load cases with axial symmetry initially spherical pores stay ellipsoids throughout the deformation history (Gărăjeu et al., 2000), whereas under compressive loading they may evolve into markedly different shapes in inelastic regimes (Segurado et al., 2002).

Cellular materials In cellular materials, such as foamed polymers or metals, wood, and trabecular bone, the volume fraction of the solid phase is low, often amounting to a few percent or less. The voids may be topologically connected (open cell foams), unconnected (closed cell foams, syntactic foams), or both connected and unconnected voids may be present (e.g., in hollow sphere foams).

Linear elastic responses are often limited to a very small range of macroscopic strains in cellular materials. Higher strains lead to nonlinear behavior in which gross shape changes of the cells take place, with bending, elastic buckling, plastic buckling, and brittle failure of cell walls or struts playing major roles on the microscale. For compression-dominated load cases this regime tends to give rise to a stress plateau on the macroscale, which underlies the favorable energy absorption properties of many cellular materials. No behavior of this type is present under tensile and shear loading. At some elevated strain the effective stiffness typically rises sharply under compression, the cellular structure having collapsed to such an extent that many cell walls or struts are in contact.

The micromechanical modeling of cellular materials has been dominated by methods based on discrete microstructures. Gibson and Ashby (1988) developed a set of well known formulae for the macroscopic behavior of cellular materials that are based on analytical models using unit cells that consist of beams (for open cell foams) and plates (for closed cell foams). In these formulae the thermomechanical moduli and other macroscopic physical properties of cellular materials are approximated by power laws in terms of the relative density.

Due to the inherent absorption contrast between matrix and voids, cellular materials are well suited to high-resolution computed tomography, giving access to microgeometries that can be directly converted into voxel models as discussed in section 3.1. This modeling strategy was first applied in bone biomechanics, where the macroscopic elastic tensors of trabecular bones have been studied by tomography-based models for more than a decade (Hollister et al., 1994; Müller and Rüegsegger, 1996). Later such models were also used for studying the nonlinear behavior of metallic and ceramic foams (Maire et al., 2000). Windowing models using periodicity compatible mixed uniform boundary conditions as discussed in section 3.3 are especially well suited for linear elastic analysis of such CT-derived microgeometries.

Unit cells for carrying out periodic homogenization of cellular materials beyond the linear elastic regime must provide for large deformations on the microscale and, typically, for mechanical instabilities of the cell walls or struts. In addition, (periodic) contact between and self contact of cell walls or struts may have to be handled. When the method of macroscopic degrees of freedom, compare section 3.2, is used, it is helpful if the vertices of the unit cell can be positioned in solid material, i.e., in cell walls or struts, so that suitable master nodes are easily available. The wavelengths of buckling modes that can be directly studied by periodic homogenization are obviously limited by the size of the unit cell. For capturing long wave buckling modes, which may well be critical, Bloch wave theory (Gong et al., 2005) can be employed.

In many cases shell elements have been used for the cell walls in FE-based discrete microstructural models of closed cell as well as hollow sphere foams, and beam elements have been employed for the struts of open cell materials. For periodic models this implies that periodicity must be enforced in terms of rotations in addition to displacements, and appropriate periodicity boundary conditions must be provided, see, e.g., Bitsche (2005). At high porosities it is necessary to account for the overlap of shell elements at edges and of beam elements at vertices when evaluating the phase volume fractions of the discretized unit cells.

The geometries of unit cells for FE-based modeling of both closed and open cell foams have typically been based on space-filling polyhedra, such as regular tetrakaidekahedra (Grenestedt, 1998), or on minimum surface shapes, such as Kelvin and Weaire–Phelan geometries (Kraynik and Reinelt, 1996). The latter geometries differ only slightly from polyhedra, but at low solid volume fractions the small distortions of the faces can lead to considerable differences in the overall mechanical response. Roberts and Garboczi (2001) proposed irregular multi-cell volume elements based on Voronoi tesselations for both closed cell and open cell foams. In addition, cell geometries based on various cubic arrangements of struts were reported for modeling scaffold-type open cell materials (Luxner, 2006).

The effects of details of the microgeometries of cellular materials, such as thickness distributions, geometrical imperfections and flaws of cell walls or struts, can considerably influence the overall behavior of cellular materials, see, e.g., Grenestedt (1998). Similar issues are raised by sintering necks in hollow sphere foams and the cross sectional shapes of open cell materials with hollow struts. Modeling such details typically requires highly resolved discretizations.

As an example fig. 14 (left) shows a Weaire–Phelan periodic model of an open cell metallic foam with hollow struts (right), which contain a second connected void region. The microgeometry was based on an idealization of the production process for such foams. First the geometry of a precursor open cell foam was obtained with the program SurfaceEvolver (Brakke, 1992) and then a coating was applied to this precursor. This coating, which represents the hollow strut foam, was discretized with continuum shell and prism-shaped solid elements provided by the FE-code ABAQUS (Simulia Corp., Pawtucket, RI). Solid volume fractions between $\xi^{(m)} = 0.01$ and

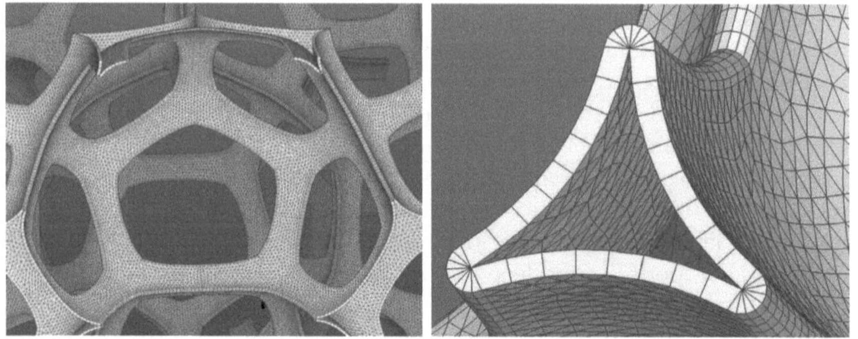

Figure 14. Weaire–Phelan model of an open cell foam with hollow struts (left) and detail of the FE-model of a strut's cross-section (right), from Daxner et al. (2007).

$\xi^{(m)} = 0.04$ were covered (Daxner et al., 2007) and elastic material behavior was specified for the solid phase. The unit cell used for modeling this cellular material is depicted in fig. 15 (left).

In the small strain regime this cellular material deforms mainly by bending of the struts and, as a consequence, the macroscopic moduli show a marked nonlinear dependence on the relative density (Daxner et al., 2007). The plot in the center of fig. 15 displays the predicted deformations of the unit cell due to a uniaxial compressive macroscopic load and that on

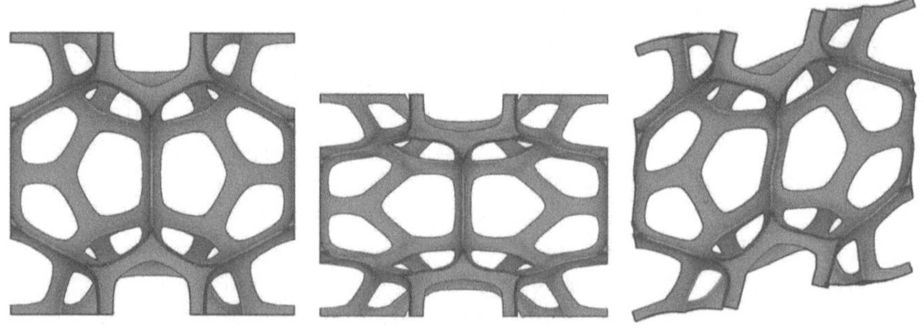

Figure 15. Undeformed state (left) and deformed configurations under compressive uniaxial (center) and shear loading (right) of the unit cell of the Weaire–Phelan model for an open cell foam with hollow struts (Daxner et al., 2007).

the right presents the deformations due to macroscopic shear loading. The cubic symmetry of the unit cell leads to a cubic elastic behavior on the macroscale, the minima and maxima of the orientation-dependent Young's moduli differing by 20% to 40%, depending on the solid volume fraction. In this complex cellular material, local contact can occur even at moderate macroscopic strains. Stability analysis of the models found both strut-level and more local instabilities on the microscale (Daxner et al., 2008).

5 Closing Remarks

Continuum micromechanics provides a considerable range of methods that differ markedly in capabilities and cost for studying the thermomechanical and thermophysical behavior of inhomogeneous materials. Current research in the field tends to emphasize, on the one hand, low cost methods that provide reasonable accuracy in describing nonlinear behavior, and, on the other hand, complex models that allow to study the stiffness, strength and conduction behavior of inhomogeneous materials at high levels of detail.

The most serious constraint to the practical use of micromechanical approaches is providing adequate material parameters for describing the in-situ behavior of the constituents, a task that is especially difficult for strength parameters.

The main strength of continuum micromechanics lies, on the one hand, in its "bidirectionality" in terms of length scales, i.e., in its capability of supporting both homogenization and localization. This allows, at least in principle, to zoom into the behavior at the microscale whenever some interesting or critical response is encountered on the structural scale. On the other hand, the input parameters into micromechanical models are neatly separated into geometrical descriptors and constituent level material parameters, which is very helpful in designing and carrying out "virtual experiments".

Continuum micromechanics has seen major development over the past decades and has been highly successful in improving the understanding of as well as in providing predictive methods for the thermomechanical and thermophysical behavior of composites, cellular materials and other inhomogeneous media.

Bibliography

J. Aboudi. The generalized method of cells and high-fidelity generalized method of cells micromechanical models — A review. *Mech. Adv. Mater. Struct.*, 11:329–366, 2004.

J. Aboudi. *Mechanics of Composite Materials*. Elsevier, Amsterdam, 1991.

J. Aboudi. Micromechanical analysis of composites by the method of cells — Update. *Appl. Mech. Rev.*, 49:83–S91, 1996.

S.G. Advani and C.L. Tucker. The use of tensors to describe and predict fiber orientation in short fiber composites. *J. Rheol.*, 31:751–784, 1987.

A. Ayyar and N. Chawla. Three-dimensional (3D) microstructure-based modeling of crack growth in particle reinforced composites. *J. Mater. Sci.*, 42:9125–9129, 2007.

Y. Benveniste. A new approach to the application of Mori–Tanaka's theory in composite materials. *Mech. Mater.*, 6:147–157, 1987a.

Y. Benveniste. Effective thermal conductivity of composites with a thermal contact resistance between the constituents: Nondilute case. *J. Appl. Phys.*, 61:2840–2843, 1987b.

Y. Benveniste and T. Miloh. The effective conductivity of composites with imperfect thermal contact at constituent interfaces. *Int. J. Engng. Sci.*, 24:1537–1552, 1986.

Y. Benveniste, G.J. Dvorak, and T. Chen. On diagonal and elastic symmetry of the approximate effective stiffness tensor of heterogeneous media. *J. Mech. Phys. Sol.*, 39:927–946, 1991.

R. Bitsche. Space filling polyhedra as mechanical models for solidified dry foams. Master's thesis, Vienna University of Technology, Vienna, Austria, 2005.

H.J. Böhm, editor. *Mechanics of Microstructured Materials*. CISM Courses and Lectures Vol. 464, Springer–Verlag, Vienna, 2004a.

H.J. Böhm. A short introduction to continuum micromechanics. In H.J. Böhm, editor, *Mechanics of Microstructured Materials*, pages 1–40, Vienna, 2004b. Springer–Verlag, CISM Courses and Lectures Vol. 464.

H.J. Böhm. A short introduction to basic aspects of continuum micromechanics. Technical Report (ILSB Arbeitsbericht 206), Vienna University of Technology, Vienna, 2007.
http://www.ilsb.tuwien.ac.at/links/downloads/ilsbrep206.pdf.

H.J. Böhm and S. Nogales. Mori–Tanaka models for the thermal conductivity of composites with interfacial resistance and particle size distributions. *Compos. Sci. Technol.*, 68:1181–1187, 2008.

H.J. Böhm, W. Han, and A. Eckschlager. Multi-inclusion unit cell studies of reinforcement stresses and particle failure in discontinuously reinforced ductile matrix composites. *Comput. Model. Engng. Sci.*, 5:5–20, 2004.

H.J. Böhm, D.H. Pahr, S. Nogales, and M.B. Gándara. Comparison of unit cell and windowing methods for obtaining estimates on macroscopic elasticity tensors of inhomogeneous materials. In Y.B. Yang, L.J. Leu, C.S. Chen, and P.C. Su, editors, *Proceedings of EASEC-11*, Taipei, 2008. National Taiwan University. paper B11–06.

M. Bornert. Homogénéisation des milieux aléatoires: bornes et estimations. In M. Bornert, T. Bretheau, and P. Gilormini, editors, *Homogénéisation en mécanique des materiaux 1. Matériaux aléatoires élastiques et milieux périodiques*, pages 133–221, Paris, 2001. Editions Hermès.

M. Bornert and P. Suquet. Propriétés non linéaires des composites: Approches par les potentiels. In M. Bornert, T. Bretheau, and P. Gilormini, editors, *Homogénéisation en mécanique des matériaux 2. Comportements non linéaires et problèmes ouverts*, pages 45–90, Paris, 2001. Editions Hermès.

M. Bornert, T. Bretheau, and P. Gilormini, editors. *Homogénéisation en mécanique des matériaux.* Editions Hermès, Paris, 2001.

K.A. Brakke. The surface evolver. *Exper. Math.*, 1:141–165, 1992.

L.M. Brown and W.M. Stobbs. The work-hardening of copper–silica. I. A model based on internal stresses, with no plastic relaxation. *Phil. Mag.*, 23:1185–1199, 1971.

J.Y. Buffière, P. Cloetens, W. Ludwig, E. Maire, and L. Salvo. In situ X-ray tomography studies of microstructural evolution combined with 3D modeling. *MRS Bull.*, 33:611–619, 2008.

V.N. Bulsara, R. Talreja, and J. Qu. Damage initiation under transverse loading of unidirectional composites with arbitrarily distributed fibers. *Compos. Sci. Technol.*, 59:673–682, 1999.

V.A. Buryachenko. The overall elastoplastic behavior of multiphase materials with isotropic components. *Acta Mech.*, 119:93–117, 1996.

J.L. Chaboche and P. Kanouté. Sur les approximations "isotrope" et "anisotrope" de l'opérateur tangent pour les méthodes tangentes incrémentale et affine. *C. R. Mecanique*, 331:857–864, 2003.

N. Chawla and K.K. Chawla. Microstructure based modeling of the deformation behavior of particle reinforced metal matrix composites. *J. Mater. Sci.*, 41:913–925, 2006.

C.M. Chimani, H.J. Böhm, and F.G. Rammerstorfer. On stress singularities at free edges of bimaterial junctions — A micromechanical study. *Scr. mater.*, 36:943–947, 1997.

P.W. Chung, K.K. Tamma, and R.R. Namburu. Asymptotic expansion homogenization for heterogeneous media: Computational issues and applications. *Composites*, 32A:1291–1301, 2001.

T. Daxner, R.D. Bitsche, and H.J. Böhm. Micromechanical models of metallic sponges with hollow struts. In T. Chandra, K. Tsuzaki, M. Militzer, and C. Ravindran, editors, *Thermec 2006*, pages 1857–1862, Trans Tech Publications, Zurich, Switzerland, 2007. Materials Science Forum 539–543.

T. Daxner, D.H. Pahr, and F.G. Rammerstorfer. Micro- and meso-instabilities in structured materials and sandwich structures. In B.G. Falzon and F.M.H. Aliabadi, editors, *Buckling and Postbuckling Structures: Experimental, Analytical and Numerical Studies*, pages 453–496, London, 2008. Imperial College Press.

I. Doghri and C. Friebel. Effective elasto-plastic properties of inclusion-reinforced composites. study of shape, orientation and cyclic response. *Mech. Mater.*, 37:45–68, 2005.

I. Doghri and A. Ouaar. Homogenization of two-phase elasto-plastic composite materials and structures. *Int. J. Sol. Struct.*, 40:1681–1712, 2003.

L. Dormieux and F.J. Ulm, editors. *Applied Micromechanics of Porous Materials.* CISM Courses and Lectures Vol. 480, Springer–Verlag, Vienna, 2005.

W.J. Drugan and J.R. Willis. A micromechanics-based nonlocal constitutive equation and estimates of representative volume element size for elastic composites. *J. Mech. Phys. Sol.*, 44:497–524, 1996.

H.L. Duan, B.L. Karihaloo, J. Wang, and X. Yi. Effective conductivities of heterogeneous media containing multiple inclusions with various spatial distributions. *Phys. Rev.*, B73:174203, 2006.

D. Duschlbauer. *Computational Simulation of the Thermal Conductivity of MMCs under Consideration of the Inclusion–Matrix Interface.* Reihe **5**, Nr.561. VDI–Verlag, Düsseldorf, 2004.

D. Duschlbauer, H.E. Pettermann, and H.J. Böhm. Mori–Tanaka based evaluation of inclusion stresses in composites with nonaligned reinforcements. *Scr. mater.*, 48:223–228, 2003.

G.J. Dvorak. Transformation field analysis of inelastic composite materials. *Proc. Roy. Soc. London*, A437:311–327, 1992.

J.D. Eshelby. The determination of the elastic field of an ellipsoidal inclusion and related problems. *Proc. Roy. Soc. London*, A241:376–396, 1957.

J.D. Eshelby. The elastic field outside an ellipsoidal inclusion. *Proc. Roy. Soc. London*, A252:561–569, 1959.

M. Ferrari. Asymmetry and the high concentration limit of the Mori–Tanaka effective medium theory. *Mech. Mater.*, 11:251–256, 1991.

J. Flaquer, A. Ríos, A. Martín-Meizoso, S. Nogales, and H.J. Böhm. Effect of diamond shapes and associated thermal boundary resistance on thermal conductivity of diamond-based composites. *Comput. Mater. Sci.*, 41: 156–163, 2007.

S.Y. Fu and B. Lauke. An analytical characterization of the anisotropy of the elastic modulus of misaligned short-fiber-reinforced polymers. *Compos. Sci. Technol.*, 58:1961–1972, 1998.

M.B. Gándara. Assessment of mixed uniform boundary conditions for predicting the macroscopic mechanical behavior of particle reinforcement composite materials. Master's thesis, Vienna University of Technology, Vienna, Austria, and Universidade de Vigo, Spain, 2008.

M. Gărăjeu, J.C. Michel, and P. Suquet. A micromechanical approach of damage in viscoplastic materials by evolution in size, shape and distribution of voids. *Comput. Meth. Appl. Mech. Engng.*, 183:223–246, 2000.

A.C. Gavazzi and D.C. Lagoudas. On the numerical evaluation of Eshelby's tensor and its application to elastoplastic fibrous composites. *Comput. Mech.*, 7:12–19, 1990.

S. Ghosh, K.H. Lee, and S. Moorthy. Two scale analysis of heterogeneous elastic-plastic materials with asymptotic homogenization and Voronoi cell finite element model. *Comput. Meth. Appl. Mech. Engng.*, 132:63–116, 1996.

L.J. Gibson and M.F. Ashby. *Cellular Solids: Structure and Properties.* Pergamon Press, Oxford, UK, 1988.

I.M. Gitman, M.B. Gitman, and H. Askes. Quantification of stochastically stable representative volumes for random heterogeneous materials. *Arch. Appl. Mech.*, 75:79–92, 2006.

I.M. Gitman, H. Askes, and L.J. Sluys. Representative volume: Existence and size determination. *Engng. Fract. Mech.*, 74:2518–2534, 2007.

M. Gologanu, J.B. Leblond, G. Perrin, and J. Devaux. Recent extensions of Gurson's model for porous ductile materials. In P. Suquet, editor, *Continuum Micromechanics*, pages 61–130, Vienna, 1997. Springer–Verlag, CISM Courses and Lectures Vol. 377.

B. Gommers, I. Verpoest, and P. van Houtte. The Mori–Tanaka method applied to textile composite materials. *Acta mater.*, 46:2223–2235, 1998.

L. Gong, S. Kyriakides, and N. Triantafyllidis. On the stability of Kelvin cell foams under compressive loads. *J. Mech. Phys. Sol.*, 53:771–794, 2005.

J.L. Grenestedt. Influence of wavy imperfections in cell walls on elastic stiffness of cellular solids. *J. Mech. Phys. Sol.*, 46:29–50, 1998.

R.E. Guldberg, S.J. Hollister, and G.T. Charras. The accuracy of digital image-based finite element models. *J. Biomech. Engng.*, 120:289–295, 1998.

A.L. Gurson. Continuum theory of ductile rupture by void nucleation and growth: Part I — Yield criteria and flow rules for porous ductile media. *J. Engng. Mater. Technol.*, 99:2–15, 1977.

Z. Hashin. Analysis of composite materials — A survey. *J. Appl. Mech.*, 50:481–505, 1983.

Z. Hashin. The differential scheme and its application to cracked materials. *J. Mech. Phys. Sol.*, 36:719–733, 1988.

Z. Hashin and B.W. Rosen. The elastic moduli of fiber-reinforced materials. *J. Appl. Mech.*, 31:223–232, 1964.

Z. Hashin and S. Shtrikman. On some variational principles in anisotropic and nonhomogeneous elasticity. *J. Mech. Phys. Sol.*, 10:335–342, 1962a.

Z. Hashin and S. Shtrikman. A variational approach to the theory of the effective magnetic permeability of multiphase materials. *J. Appl. Phys.*, 33:3125–3131, 1962b.

B. Hassani and E. Hinton. *Homogenization and Structural Topology Optimization.* Springer–Verlag. London, 1999.

D.P.H. Hasselman and L.F. Johnson. Effective thermal conductivity of composites with interfacial thermal barrier resistance. *J. Compos. Mater.*, 21:508–515, 1987.

H. Hatta and M. Taya. Equivalent inclusion method for steady state heat conduction in composites. *Int. J. Engng. Sci.*, 24:1159–1172, 1986.

S. Hazanov. Hill condition and overall properties of composites. *Arch. Appl. Mech.*, 68:385–394, 1998.

S. Hazanov and M. Amieur. On overall properties of elastic bodies smaller than the representative volume. *Int. J. Engng. Sci.*, 33:1289–1301, 1995.

Q. He and A. Curnier. A more fundamental approach to damage elastic stress–strain relations. *Int. J. Sol. Struct.*, 32:1433–1457, 1997.

R. Hill. The elastic behavior of a crystalline aggregate. *Proc. Phys. Soc. London*, A65:349–354, 1952.

R. Hill. Elastic properties of reinforced solids: Some theoretical principles. *J. Mech. Phys. Sol.*, 11:357–372, 1963.

R. Hill. A self-consistent mechanics of composite materials. *J. Mech. Phys. Sol.*, 13:213–222, 1965a.

R. Hill. Continuum micro-mechanics of elastic-plastic polycrystals. *J. Mech. Phys. Sol.*, 13:89–101, 1965b.

R. Hill. The essential structure of constitutive laws for metal composites and polycrystals. *J. Mech. Phys. Sol.*, 15:79–95, 1967.

S.J. Hollister, J.M. Brennan, and N. Kikuchi. A homogenization sampling procedure for calculating trabecular bone effective stiffness and tissue level stress. *J. Biomech.*, 27:433–444, 1994.

C.O. Huber. *Numerical Simulations of Metal Matrix Composites — Tribological Behavior and Finite Strain Response on Different Length Scales.* PhD thesis, Vienna University of Technology, Vienna, Austria, 2008.

C.O. Huber, M.H. Luxner, S. Kremmer, S. Nogales, H.J. Böhm, and H.E. Pettermann. Forming simulations of MMC components by a micromechanics based hierarchical FEM approach. In J.M.A.C. de Sa and A.D. Santos, editors, *Proceedings of NUMIFORM 2007*, pages 1351–1356, New York, NY, 2007. American Institute of Physics.

C. Huet. Coupled size and boundary-condition effects in viscoelastic heterogeneous and composite bodies. *Mech. Mater.*, 31:787–829, 1999.

S. Jansson. Homogenized nonlinear constitutive properties and local stress concentrations for composites with periodic internal structure. *Int. J. Sol. Struct.*, 29:2181–2200, 1992.

D. Jeulin. Random structure models for homogenization and fracture statistics. In D. Jeulin and M. Ostoja-Starzewski, editors, *Mechanics of Random and Multiscale Microstructures*, pages 33–91, Vienna, 2001. Springer–Verlag, CISM Courses and Lectures Vol. 430.

M. Jiang, M. Ostoja-Starzewski, and I. Jasiuk. Scale-dependent bounds on effective elastoplastic response of random composites. *J. Mech. Phys. Sol.*, 49:655–673, 2001.

M. Jiang, I. Jasiuk, and M. Ostoja-Starzewski. Apparent thermal conductivity of periodic two-dimensional composites. *Comput. Mater. Sci.*, 25: 329–338, 2002a.

M. Jiang, M. Ostoja-Starzewski, and I. Jasiuk. Apparent elastic and elastoplastic behavior of periodic composites. *Int. J. Sol. Struct.*, 39:199–212, 2002b.

J.W. Ju and L.Z. Sun. Effective elastoplastic behavior of metal matrix composites containing randomly located aligned spheroidal inhomogeneities. Part I: Micromechanics-based formulation. *Int. J. Sol. Struct.*, 38:183–201, 2001.

J.W. Ju and L.Z. Sun. A novel formulation for the exterior point Eshelby's tensor of an ellipsoidal inclusion. *J. Appl. Mech.*, 66:570–574, 1999.

M. Kachanov, I. Tsukrov, and B. Shafiro. Effective moduli of solids with cavities of various shapes. *Appl. Mech. Rev.*, 47:151–S174, 1994.

M. Kailasam, N. Aravas, and P. Ponte Castañeda. Porous metals with developing anisotropy: Constitutive models, computational issues and applications to deformation processing. *Comput. Model. Engng. Sci.*, 1: 105–118, 2000.

T. Kanit, S. Forest, I. Gallier, V. Mounoury, and D. Jeulin. Determination of the size of the representative volume element for random composites: Statistical and numerical approach. *Int. J. Sol. Struct.*, 40:3647–3679, 2003.

Z.F. Khisaeva and M. Ostoja-Starzewski. On the size of RVE in finite elasticity of random composites. *J. Elast.*, 85:153–173, 2006.

V.G. Kouznetsova, M.G.D. Geers, and W.A.M. Brekelmans. Multi-scale constitutive modeling of heterogeneous materials with a gradient-enhanced computational homogenization scheme. *Int. J. Num. Meth. Engng.*, 54:1235–1260, 2002.

A.M. Kraynik and D.A. Reinelt. The linear elastic behavior of a bidisperse weaire–phelan soap foam. *Chem. Engng. Comm.*, 150:409–420, 1996.

H. Ledbetter. Monocrystal–polycrystal elastic constants of a stainless steel. *Phys. Stat. Sol. A*, 85:89–96, 1984.

V.M. Levin. On the coefficients of thermal expansion of heterogeneous materials. *Mech. Sol.*, 2:58–61, 1967.

M.H. Luxner. *Modeling and Simulation of Highly Porous Open Cell Structures — Elasto-Plasticity and Localization versus Disorder and Defects.* Reihe **18**, Nr.308. VDI–Verlag, Düsseldorf, 2006.

E. Maire, F. Wattebled, J.Y. Buffière, and G. Peix. Deformation of a metallic foam studied by X-ray computed tomography and finite element calculations. In T.W. Clyne and F. Simancik, editors, *Metal Matrix Composites and Metallic Foams*, pages 68–73, Weinheim, 2000. Wiley–VCH.

K.Z. Markov and L. Preziosi. *Heterogeneous Media: Micromechanics Modeling Methods and Simulations.* Birkhäuser, Boston, MA, 2000.

R. Masson, M. Bornert, P. Suquet, and A. Zaoui. An affine formulation for the prediction of the effective properties of nonlinear composites and polycrystals. *J. Mech. Phys. Sol.*, 48:1203–1227, 2000.

C.F. Matt and M.A.E. Cruz. Effective thermal conductivity of composite materials with 3-D microstructures and interfacial thermal resistance. *Numer. Heat Transf.*, A53:577–604, 2008.

R. McLaughlin. A study of the differential scheme for composite materials. *Int. J. Engng. Sci.*, 15:237–244, 1977.

A.R. Melro, P.P. Camanho, and S.T. Pinho. Generation of random distribution of fibres in long-fibre reinforced composite. *Compos. Sci. Technol.*, 68:2092–2102, 2008.

J.C. Michel and P. Suquet. Nonuniform transformation field analysis. *Int. J. Sol. Struct.*, 40:6937–6955, 2003.

J.C. Michel, H. Moulinec, and P. Suquet. Effective properties of composite materials with periodic microstructure: A computational approach. *Comput. Meth. Appl. Mech. Engng.*, 172:109–143, 1999.

C.A. Miller and S. Torquato. Effective conductivity of hard sphere suspensions. *J. Appl. Phys.*, 68:5486–5493, 1990.

G.W. Milton. *The Theory of Composites.* Cambridge University Press, Cambridge, 2002.

G.W. Milton and R.V. Kohn. Variational bounds on the effective moduli of anisotropic composites. *J. Mech. Phys. Sol.*, 36:597–629, 1988.

A. Molinari, G.R. Canova, and S. Ahzi. A self-consistent approach for large deformation viscoplasticity. *Acta metall.*, 35:2983–2984, 1987.

T. Mori and K. Tanaka. Average stress in the matrix and average elastic energy of materials with misfitting inclusions. *Acta metall.*, 21:571–574, 1973.

R. Müller and P. Rüegsegger. Analysis of mechanical properties of cancellous bone under conditions of simulated bone athropy. *J. Biomech.*, 29:1053–1060, 1996.

T. Mura. *Micromechanics of Defects in Solids.* Martinus Nijhoff, Dordrecht, 1987.

S. Nemat-Nasser and M. Hori. *Micromechanics: Overall Properties of Heterogeneous Solids.* North–Holland, Amsterdam, 1993.

S. Nogales. *Numerical Simulation of the Thermal and Thermomechanical Behavior of Metal Matrix Composites.* Reihe **18**, Nr.317. VDI–Verlag, Düsseldorf, 2008.

S. Nogales and H.J. Böhm. Modeling of the thermal conductivity and thermomechanical behavior of diamond reinforced composites. *Int. J. Engng. Sci.*, 46:606–619, 2008.

J.F. Nye. *Physical Properties of Crystals, Their Representation by Tensors and Matrices.* Clarendon, Oxford, UK, 1957.

M. Ostoja-Starzewski. Material spatial randomness: From statistical to representative volume element. *Probab. Engng. Mech.*, 21:112–131, 2006.

M. Ostoja-Starzewski. Random field models of heterogeneous materials. *Int.J.Sol.Struct.*, 35:2429–2455, 1998.

D.H. Pahr and H.J. Böhm. Assessment of mixed uniform boundary conditions for predicting the mechanical behavior of elastic and inelastic discontinuously reinforced composites. *Comput. Model. Engng. Sci.*, 34: 117–136, 2008.

D.H. Pahr and P.K. Zysset. Influence of boundary conditions on computed apparent elastic properties of cancellous bone. *Biomech. Model. Mechanobiol.*, 7:463–476, 2008.

S.D. Papka and S. Kyriakides. In-plane biaxial crushing of honeycombs — Part II: Analysis. *Int. J. Sol. Struct*, 36:4397–4423, 1999.

H.E. Pettermann. *Derivation and Finite Element Implementation of Constitutive Material Laws for Multiphase Composites Based on Mori–Tanaka Approaches.* Reihe **18**, Nr.217. VDI–Verlag, Düsseldorf, 1997.

H.E. Pettermann, H.J. Böhm, and F.G. Rammerstorfer. Some direction dependent properties of matrix–inclusion type composites with given reinforcement orientation distributions. *Composites B*, 28B:253–265, 1997.

P. Ponte Castañeda. Bounds and estimates for the properties on nonlinear inhomogeneous systems. *Phil. Trans. Roy. Soc.*, A340:531–567, 1992.

P. Ponte Castañeda and P. Suquet. Nonlinear composites. In E. van der Giessen and T.Y. Wu, editors, *Advances in Applied Mechanics 34*, pages 171–302, New York, NY, 1998. Academic Press.

P. Ponte Castañeda and J.R. Willis. The effect of spatial distribution on the effective behavior of composite materials and cracked media. *J. Mech. Phys. Sol.*, 43:1919–1951, 1995.

J. Qu and M. Cherkaoui. *Fundamentals of Micromechanics of Solids.* John Wiley, New York, NY, 2006.

A. Reuss. Berechnung der Fließgrenze von Mischkristallen auf Grund der Plastizitätsbedingung für Einkristalle. *ZAMM*, 9:49–58, 1929.

M. Rintoul and S. Torquato. Reconstruction of the structure of dispersions. *J. Colloid Interf. Sci.*, 186:467–476, 1997.

A.P. Roberts and E.J. Garboczi. Elastic moduli of model random three-dimensional closed-cell cellular solids. *Acta mater.*, 49:189–197, 2001.

P.W. Ruch, O. Beffort, S. Kleiner, L. Weber, and P.J. Uggowitzer. Selective interfacial bonding in Al(Si)–diamond composites and its effect on thermal conductivity. *Compos. Sci. Technol.*, 66:2677–2685, 2006.

J. Schjødt-Thomsen and R. Pyrz. The Mori–Tanaka stiffness tensor: Diagonal symmetry, complex fibre orientations and non-dilute volume fractions. *Mech. Mater.*, 33:531–544, 2001.

J. Segurado. *Micromecánica computacional de materiales compuestos reforzados con partículas.* PhD thesis, Universidad Politécnica de Madrid, Spain, 2004.

J. Segurado, E. Parteder, A. Plankensteiner, and H.J. Böhm. Micromechanical studies of the densification of porous molybdenum. *Mater. Sci. Engng.*, A333:270–278, 2002.

T. Siegmund, R. Cipra, J. Liakus, B. Wang, M. LaForest, and A. Fatz. Processing–microstructure–property relationships in short fiber reinforced carbon–carbon composite system. In H.J. Böhm, editor, *Mechanics of Microstructured Materials*, pages 235–258, Vienna, 2004. Springer–Verlag, CISM Courses and Lectures Vol. 464.

R.J.M. Smit, W.A.M. Brekelmans, and H.E.H. Meijer. Prediction of the mechanical behavior of non-linear heterogeneous systems by multi-level finite element modeling. *Comput. Meth. Appl. Mech. Engng.*, 155:181–192, 1998.

M. Stroeven, H. Askes, and L.J. Sluys. Numerical determination of representative volumes for granular materials. *Comput. Meth. Appl. Mech. Engng.*, 193:3221–3238, 2004.

L.Z. Sun, J.W. Ju, and H.T. Liu. Elastoplastic modeling of metal matrix composites with evolutionary particle debonding. *Mech. Mater.*, 35:559–569, 2003.

P. Suquet, editor. *Continuum Micromechanics.* CISM Courses and Lectures Vol. 377, Springer–Verlag, Vienna, 1997.

S. Swaminathan, S. Ghosh, and N.J. Pagano. Statistically equivalent representative volume elements for unidirectional composite microstructures: Part I — Without damage. *J. Compos. Mater.*, 40:583–604, 2006.

G.P. Tandon and G.J. Weng. A theory of particle-reinforced plasticity. *J. Appl. Mech.*, 55:126–135, 1988.

S. Torquato. *Random Heterogeneous Media.* Springer–Verlag. New York, NY, 2002.

S. Torquato. Morphology and effective properties of disordered heterogeneous media. *Int. J. Sol. Struct.*, 35:2385–2406, 1998a.

S. Torquato. Effective stiffness tensor of composite media: II. Applications to isotropic dispersions. *J. Mech. Phys. Sol.*, 46:1411–1440, 1998b.

S. Torquato and D.M. Rintoul. Effect of the interface on the properties of composite media. *Phys. Rev. Lett.*, 75:4067–4070, 1995.

S. Torquato, F. Lado, and P.A. Smith. Bulk properties of two-phase disordered media. IV. Mechanical properties of suspensions of penetrable spheres at nondilute concentrations. *J. Chem. Phys.*, 86:6388–6392, 1987.

D. Trias, J. Costa, A. Turon, and J.F. Hurtado. Determination of the critical size of a statistical representative volume element (SRVE) for carbon reinforced polymers. *Acta mater.*, 54:3471–3484, 2006.

E. van der Giessen and V. Tvergaard. Development of final creep failure in polycrystalline aggregates. *Acta metall. mater.*, 42:952–973, 1994.

W. Voigt. Über die Beziehung zwischen den beiden Elasticitäts-Constanten isotroper Körper. *Ann. Phys.*, 38:573–587, 1889.

G.J. Weng. The theoretical connection between Mori–Tanaka theory and the Hashin–Shtrikman–Walpole bounds. *Int. J. Engng. Sci.*, 28:1111–1120, 1990.

O. Wiener. Die Theorie des Mischkörpers für das Feld der stationären Strömung. *Abh. Math.-Phys. Kl. Königl. Sächs. Ges. Wiss.*, 32:509–604, 1912.

J.R. Willis. Bounds and self-consistent estimates for the overall moduli of anisotropic composites. *J. Mech. Phys. Sol.*, 25:185–202, 1977.

P.J. Withers. The determination of the elastic field of an ellipsoidal inclusion in a transversely isotropic medium, and its relevance to composite materials. *Phil. Mag.*, A59:759–781, 1989.

A. Zaoui. Continuum micromechanics: Survey. *J. Engng. Mech.*, 128:808–816, 2002.

J. Zeman. *Analysis of Composite Materials with Random Microstructure.* PhD thesis, Czech Technical University, Prague, Czech Republic, 2003.

J. Zeman and M. Šejnoha. Numerical evaluation of effective elastic properties of graphite fiber tow impregnated by polymer matrix. *J. Mech. Phys. Sol.*, 49:69–90, 2001.

T.I. Zohdi and P. Wriggers. A model for simulating the deterioration of structural-scale material responses of microheterogeneous solids. *Comput. Meth. Appl. Mech. Engng.*, 190:2803–2823, 2001.

Effect of Microstructure: Multi-scale Modelling

Vadim V. Silberschmidt

Wolfson School of Mechanical and Manufacturing Engineering,
Loughborough University, UK

1 Introduction

A study of a structure of any real material would vividly demonstrate the presence of several hierarchical levels – from atoms to microscopic features to the macroscopic scale of a specimen/component/structure. Depending on the type of material and its microstructure, several scales – six for highly heterogeneous cases such as composite structures (Beaumont *et al.*, 2008) and bones (Rho *et al.*, 1998) – can be introduced. These scales can cover in various materials the range of lengths from nanometres to centimetres or metres. An obvious reason for introduction of any additional scale into consideration is specificity of its structure and/or of the character of realisation of deformation and failure processes at the corresponding scale.

The presence of such scales presupposes an introduction of scale-specific constitutive descriptions as well as conditions of the transfer from one scale to another, i.e. the effect of processes at one scale on properties of, or processes at, another one. A growing understanding of necessity for adequate mechanical descriptions of real materials with hierarchical structures has been a reason of an active research in this area (see, e.g., recent books by Soutis and Beaumont (2005) and Kwon *et al.* (2008) on composites). And various models and descriptions of deformation and fracture processes have been developed for different scales. First-principle methods could be applied at the lowest scales but still, in words of David McDowell (2008), 'bottom–up modelling of realistic multiphase, hierarchical microstructures remains an outstanding grand challenge'.

In practical terms it means that, currently, applications of modelling tools do not cover *all* the scales that are present in the studied material; rather, they are focused on the scale of interest and adjacent ones: an account for the directly underlying scale allows to incorporate finer details of the structure while the overlying scale is used to define the long-range effects and/or loading/environmental conditions as well as geometric or kinematic constraints. Obviously, such a reduction of the number of analysed scales diminishes the complexity of the problem.

If the study is limited to analysis of the material's effective properties – in contrast to the study of a component with exact geometry and loading – complexity of a modelling approach can be further diminished: an effect of the overlying scale can be reduced to periodic boundary conditions imposed onto a *representative volume element* (RVE) or *unit cell*. Such a transition is equivalent to an assumption that a real material can be presented as consisting of similar elements with the same properties/behaviour. Notwithstanding obvious advantages of such an approach – the main being a refinement of presentation of the structure *within* a RVE – it has some limitations due to finiteness of RVE's dimensions. Another problem is linked with a way to introduce a microstructure into the RVE. The use of the scanned data for a real microstructure and its meshing by finite elements is an option preferred by many. Unfortunately, it leaves unanswered the question of representativeness of the chosen image. A use of 'artificial', or numerically generated, microstructures that have the same parameters as the real microstructure (see Silberschmidt (2008) for their review) is another approach, and a rather cumbersome one. In other words, the powerful tool of RVEs is not universally applicable (the next Section will give more details on this).

This Chapter does not give an exhausting review of all classes of materials or modelling tools and strategies. It covers some aspects of multi-scale modelling of microstructured materials and presents several schemes of introduction of microstructural features into numerical simulations in order to provide a more adequate description of various properties of such materials or processes in them. The discussion concentrates on the problems where a scale of several micrometres is important; thus descriptions of lower scales (e.g. atomic) are not treated here. A seemingly arbitrary choice of materials and applications is linked to research interests of the author.

2. Account for microstructural randomness of composites

Traditional composite materials, for instance, carbon fibre-reinforced polymer (CFRP) laminates usually have a rather random microstructure due to manufacturing technologies that cannot provide a spatial periodicity of reinforcing elements. As a result, images of their microstructures demonstrate a rather random pattern of distribution of constituents. For instance, the pattern of fibres in a digitally enhanced image of a unidirectional carbon-fibre reinforced

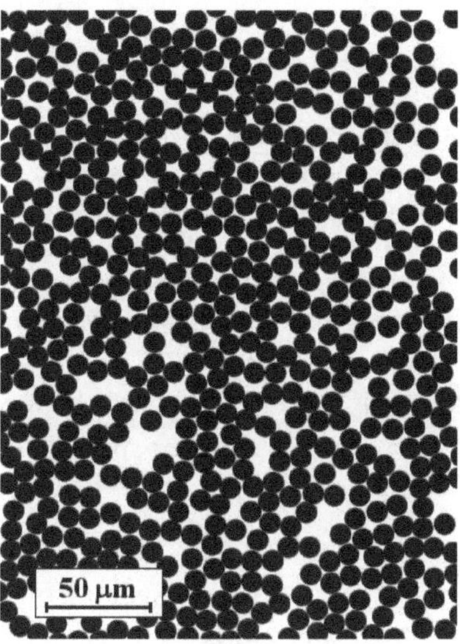

Figure 1. Distribution of continuous graphite fibres in epoxy matrix in a transverse cross section of a unidirectionally reinforced ply (digitalisation of a micrograph).

composite in Figure 1 can hardly be reduced to a periodic – square, hexagonal etc. – one. As a result, dimensions of RVEs, used to model the properties of this material, should be larger than an (average) distance between the centres of neighbouring fibres. A detailed analysis of various parameters of the set of the fibres is given elsewhere (Silberschmidt, 2005a, 2006), here the main implications for numerical simulations are discussed.

From Mechanics of Composites (see, e.g. Herakovich (1988)) we know that longitudinal stiffness of the unidirectional fibrous composite is very close to the one defined by the linear rule of mixture and, hence, proportional to the volume fraction of reinforcement:

$$\overline{E}_{11} \approx E_{11}^{\mathrm{RoM}} = v^{\mathrm{f}}\left(E_{\mathrm{L}}^{\mathrm{f}} - E^{\mathrm{m}}\right) + E^{\mathrm{m}}, \tag{1}$$

where $\overline{E}_{11}$ is the effective axial modulus of the composite, E_{11}^{RoM} is the axial modulus calculated according to the linear rule of mixtures, $E_{\mathrm{L}}^{\mathrm{f}}$ is the longitudinal modulus of fibres (transversely isotropic in the case of carbon ones) and E^{m} is the Young's modulus of the (isotropic) matrix. Equation (1) presents the effective magnitude for the composite, i.e. the one, averaged over a large volume.

Still, from Figure 1 it is apparent that *local volume fractions* differ for various parts of the presented cross section so one can calculate the local magnitudes of the Young's modulus. Obviously, moving a window of specific size to select a local area of the cross section, one would get changing magnitudes of this parameter for various window sizes. This data, presented as histograms (Figure 2), demonstrates that the decrease in the window size results in a larger scatter in the local volume fractions of fibres. This can seriously affect the local magnitudes of E_{11}^{RoM}, especially in materials with a high contrast of properties of their constituents; CFRP is a good example of such materials, with the Young's modulus of matrix being only a small fraction of the is the longitudinal modulus of fibres – 2.3%.

Figure 3 presents the extent of fluctuations of the local magnitudes of the effective longitudinal modulus of unidirectional CFRP normalised by its global value (i.e. for an infinitely large area). It is seen that even at the scale of 125 μm that corresponds to a standard ply thickness in CFRPs the obtained maximum and minimum values of modulus do not converge entirely.

A large scatter at the low scale corresponds, on the one hand, to the presence of resin-rich areas practically without fibres (lower bound) and, on the other hand, areas occupied by a single fibre (upper bound). This has direct implications for numerical simulations at the microscopic scale – an increase in the number of elements (mesh refinement) in such simulations would not improve the quality of obtained results if it is not accompanied by the introduction of statistic tools to accommodate material's heterogeneity.

As was discussed above, the study of effective properties and performance of heterogeneous materials would generally not be seriously affected if standard FE approaches based on global properties of (anisotropic) layers are used. In contrast, analysis of fracture processes – crack formation, initiation and propagation of delamination zones, etc. – that are spatially localised should contain a direct account for microstructural randomness. Let us consider, for instance, a study of cracking in CFRP laminates. In order to properly account for the through-thickness properties in plies the thickness of each ply should be divided into at least 4 elements with a size of some 30 μm. For such meshing the scatter in local levels of the Young's modulus, measured as its ratio of maximum magnitude to the minimum one, would be more than 200% (see Figure 3). And refining the area would only increase this scatter!

A direct consequence of this scatter is a highly non-uniform distribution of stresses in CFRPs even under uniform macroscopic loading conditions, e.g. tensile fatigue that is one of the main test techniques in analysis of the damage

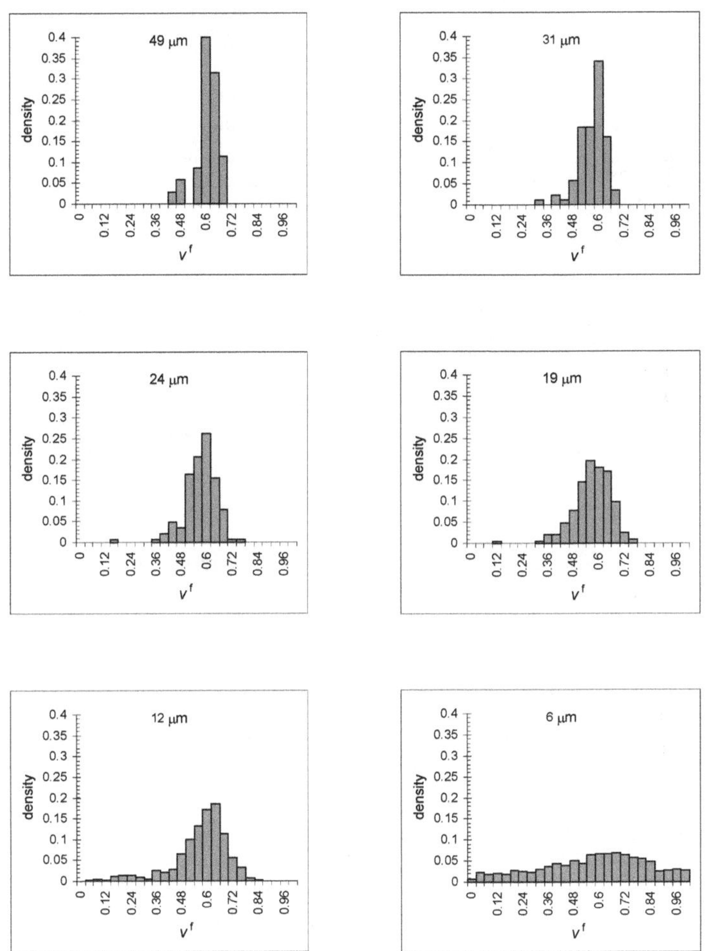

Figure 2. Effect of the length scale (window size) on distribution of the volume fraction of fibres.

tolerance of laminates used in aerospace applications. The stress localisation is affected not only by fluctuations of the local volume fraction of reinforcement but also by the character of fibres closeness to each other.

Figure 1 vividly demonstrates that some neighbouring fibres touch each other while other fibres are separated from their neighbours by a considerable amount

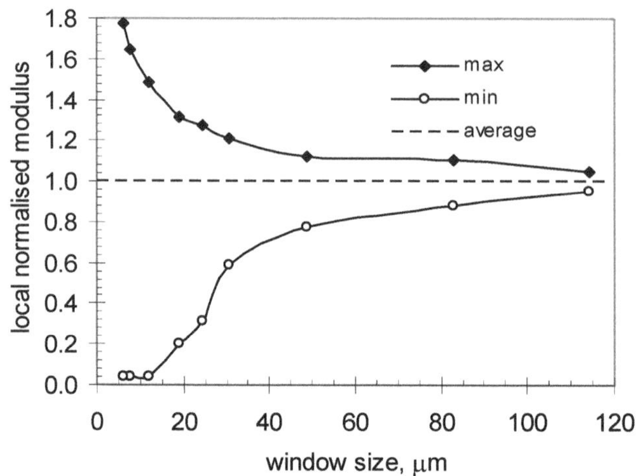

Figure 3. Effect of the length scale on bounds of distributions of the normalised axial modulus.

of matrix. A quantitative data for spacings between nearest neighbours in the studied ensemble is presented in Figure 4. Narrow areas between fibres, especially between touching ones, are linked to local stress concentration and can serve as crack generation sites.

Such load transfer mechanisms – from macroscopically uniform applied loads to highly localised stress concentrations – are responsible for spatially non-uniform damage evolution in laminates. One of the obvious examples is the character of matrix cracking in $[0_m/90_n]_s$ cross-ply CFRP laminates Silberschmidt, 1995). Matrix (transversal) cracks appear in weak 90° layers of the sandwich-like cross-ply laminates at the very early stages of loading, for instance, they can be initiated during a few initial cycles of tensile fatigue. They propagate in a weak matrix parallel to strong fibres (and nearly orthogonal to the applied tension) not causing their fracture. The density of matrix cracks increases (an average spacing between neighbouring cracks diminishes) with the loading history. Since they cross the entire thickness of 90° plies they cause the so called *shielding effect* – decreasing the axial stress component in the area in their direct vicinity (Silberschmidt, 1998). As a result, the increase in the numbers of transverse cracks decelerates at the advanced stages of fatigue reaching a (nearly) constant density known as *characteristic damage state* (Reifsnider and

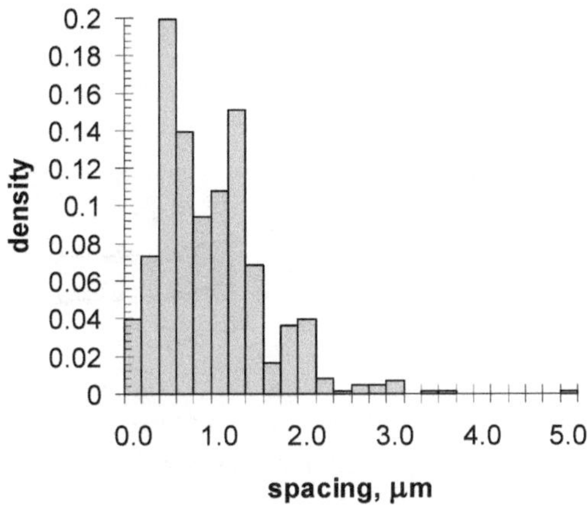

Figure 4. Distribution of spacing between the nearest neighbours in the set of carbon fibres in unidirectional laminate.

Case, 2002).

This type of behaviour caused various researchers to introduce the following model scenario for matrix cracking: the first crack appears in the middle of the specimen, the next generation – in the mid-spacing between it and the edges etc. As a result, each stage of the cracking process is characterised by a periodic pattern of matrix cracks with a constant spacing between neighbours equal to $L/(n+1)$, where L is the axial length of the laminate and n is a number of matrix crack in its 90° layer.

In real laminates, matrix cracking is anything but an ordered process. Figure 5 demonstrates the change in the parameters of the spacing distribution of matrix cracks in the $[0_4/90_4/0_4]$ CFRP laminate with the loading history, obtained from the results of X-ray radiography analysis. It is obvious that though the increase in the number of cycles causes a slow down in the decline of crack spacing, the distribution of matrix cracks is highly non-uniform. Even after 4 million cycles the ratio of spacings corresponding to 0.95 and 0.05 quantile is more than 3.1.

This data contravene a major assumption of numerical schemes traditionally used to model matrix cracking in laminates – linking the axial size of RVEs with the average spacing between neighbouring transverse cracks. In (Silberschmidt, 2005b) a direct investigation of this assumption demonstrated its limitation. Four

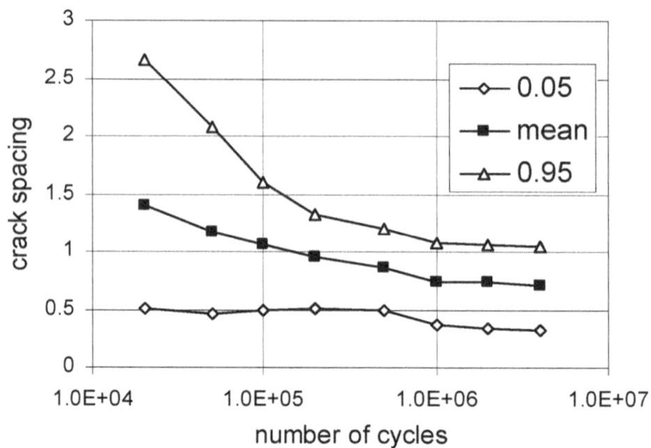

Figure 5. Evolution of parameters of spacing distribution – mean value as well as 0.05 and 0.95 quantiles – in $[0_4/90_4/0_4]$ CFRP laminate.

various distributions of matrix cracks were studied: a random one, based on the experimental data, and three artificial distributions with uniform spacings, corresponding to the minimum, maximum and mean spacings from the experimental data. In this study the ratio of the maximum spacing to the minimum one was 4.0 (the spacing magnitudes correspond to 0.05 and 0.95 quantiles of $[0_1/90_4/0_1]$ laminate). Two distributions – the random one and the one with the minimum spacing $a_{\min} = 0.4$ mm contained at least one pair of neighbouring matrix cracks separated by this distance. The difference was that in the former case other spacings differs form that spacing while in the latter case they were all equal to it.

The finite-element simulation using the effective properties for lamina in order to study a pure effect were performed; the obtained results for the axial stress in 0° layers (near the 0/90 interface) are given in Figure 6. The difference for two sets of cracks demonstrates that the use of RVEs (i.e. a spatially uniform distribution of cracks) results in underestimation of the extent of the stress concentration near the tips of matrix cracks at the interface between week (90°) and strong (0°) layers.

To overcome the limitations of the RVE-based approaches, various schemes can be used. A lattice model based on the direct introduction of the material randomness in terms of its local stiffness (linked to spatial fluctuations of the

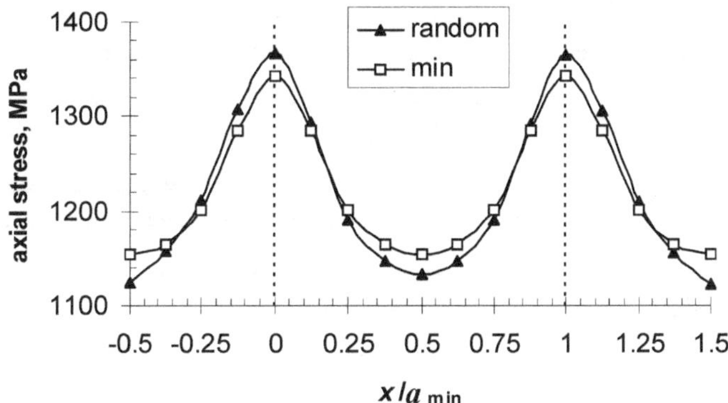

Figure 6. Distributions of axial stress in 0° layers (near 0/90 interface) for spacings of the same length in laminates with two different sets of matrix cracks.

volume fraction of fibres) was suggested in (Silberschmidt, 1997). The load transition mechanism, developed there, is a result of interaction between ordering factors (e.g. due to the shielding zones near cracks) and random initial distributions of weak and strong areas. The model also accounts for damage evolution in elements with random properties, forming a lattice, and a local fracture criterion in terms of the critical damage level. As a result, it allows describing a process of matrix cracking that reminds a real one (see Figure 7).

The crack sets at various stages of loading history are characterised by a changing degree of randomness – more pronounced at initial stages of life and less – at the advanced stages of tensile fatigue. At the latter stages, newly-formed transverse cracks (shown by thick lines in Figure 7) tend to form in mid-spacing areas. This can be naturally explained by the mutual action of the shielding zones from the neighbouring cracks; the total length of these zones at this stage exceeds the average spacing. Still, at all stages there are cracks that are formed relatively close to already existing ones (see Figures 7b and 7d). In such cases, the effect of large local fluctuations in material's properties is strong enough to compensate for the local stress reduction due to the shielding effect that can not prevent generation of a new matrix crack.

Some other models incorporating random material properties of laminates are discussed, for instance, in (Berthelot, 2003). A recently developed approach is based on transition to *stochastic cohesive zone elements* (Khokhar *et al.*, 2008)

Figure 7. Positions of matrix cracks in a specimen (axial length 50 mm) of $[0_2/90_8/0_2]$ T300-934 laminate loaded by tensile fatigue with the maximum cyclic stress 450 MPa at different moments of loading history: (a) 100 cycles, (b) 4×10^3 cycles, (c) 10^5 cycles and (d) 2×10^5 cycles. Vertical to horizontal scale 2:1,

(as opposed to standard ones). In this approach, parameters of the traction-separation law for cohesive zone elements are random magnitudes in order to reflect local differences in their microstructures.

3. Modelling ceramic coatings with random porosity

Ceramic coatings are widely used to protect components and structures from wear, chemical attacks and high-temperature environments. Various deposition processes are used to manufacture, for instance, Thermal Barrier Coatings (TBCs); thermal plasma spraying is one of the most common processes for high-performance coatings. In this process molten droplets, formed from ceramic powder particles, are deposited at high temperature and velocity on a cooler metallic substrate and rapidly cooled. After deposition, the coating is heated to fuse the material into a dense alloyed structure and produce a diffusion bond to the substrate. A layer of coating is built up by moving the plasma-gun, spraying droplets, transversely across the substrate.

Such manufacturing processes of deposition define a specific microstructure of plasma sprayed ceramic TBCs – they have porous lamellar microstructures consisting of elongated and flat splats. A typical microstructure of alumina (Al_2O_3) coating is presented in Figure 8. This microstructure greatly influences

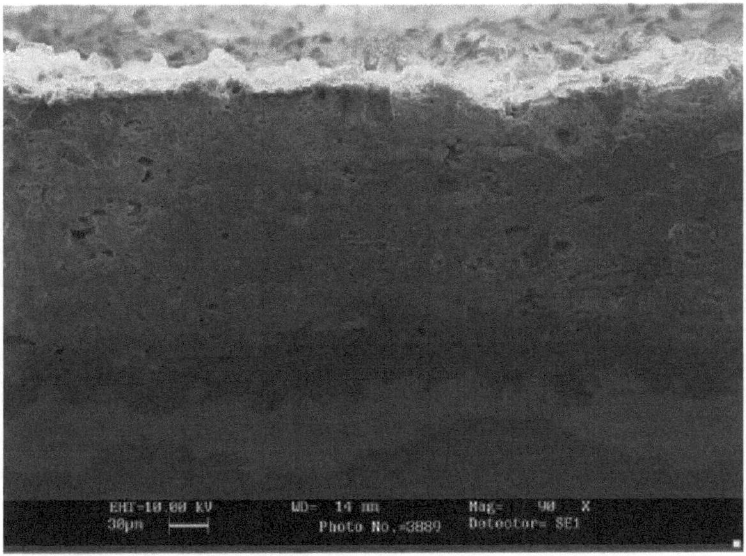

Figure 8. SEM cross-sectional micrograph of alumina coating on a substrate.

effective mechanical properties of coatings, resulting in their pronounced anisotropy as well as a significant – up to 75% – reduction in their stiffness as compared to that for the respective bulk material. This reduction increases with the level of porosity. The anisotropy ratio $C_{22,33}/C_{11}$ (axis 1 corresponding to deposition direction while 2 and 3 to directions in the deposition plane) of the as-sprayed alumina ranges from 1.8 to 3.6 depending on the material deposition process (Wanner and Lutz, 1998). There is no significant difference between the in-plane properties, i.e. it could be considered that $C_{22} = C_{33}$ (Damani and Wanner, 2000) and the ceramic coating is transversely isotropic.

As in composites, with the standard thickness of coating being 200 μm – 500 μm and the void size up to 50 μm, analysis of the damage evolution and cracking under various loading and/or environmental conditions needs an introduction of microstructure-related properties. In standard approaches, porosity is used as the main – and in many cases single – parameter together with introduction of empirical relationships for deterioration of the Young's modulus with the increase in the porosity level. Such an approach does not account for significant differences in the shape and size of voids observed in alumina coatings with the latter additionally affecting anisotropy of their properties.

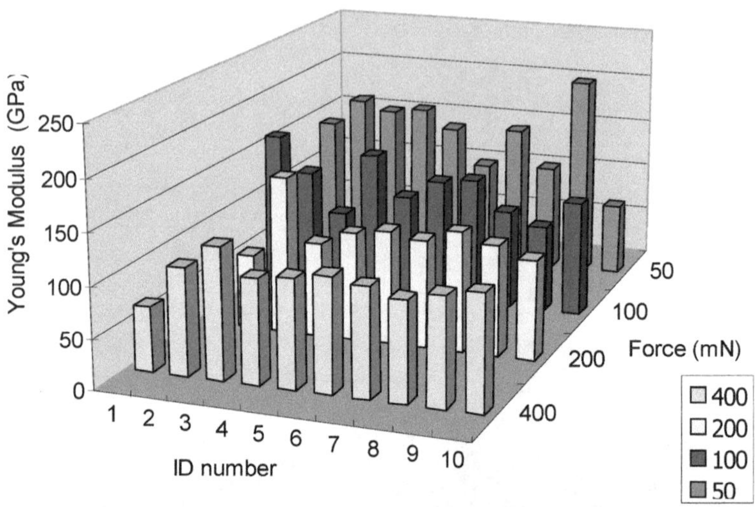

Figure 9. Values of Young's modulus of alumina coating in the deposition direction from nanoindentation tests (distance between neighbouring points 100 μm).

Experimental results, based on nanoindentation, show significant fluctuation in the local magnitudes of the Young's modulus of alumina coating (Figure 9) demonstrating the necessity to transfer from the effective (global) properties to microstructure-defined local ones.

So, this Section presents an alternative approach, based on a multi-scale modelling scheme. The details of the approach are presented elsewhere (Zhao and Silberschmidt, 2005, 2006 and Zhao, 2005).

3.1. Modelling microstructure of ceramic coatings

The first step to introduce the microstructure and its main features into the model of the alumina coating is to quantify respective parameters. The results of research into *i*-th damage evolution in bulk alumina (Najar and Silberschmidt, 1998) demonstrated that porosity is the dominant factor affecting this process. Microvoids exist in coatings at random locations and their interaction is neglected for the considered porosity level (below 10%) on the assumption that the characteristic interaction length is smaller than dimensions of elements and spacing between the neighbouring microvoids. To characterise porosity in ceramic coatings standard parameters are used:

Porosity p in ceramic coatings is generally defined as a volume or an area fraction of microvoids:

$$p = \frac{1}{V}\sum_{i=1}^{n} V_i \quad \text{or} \quad p = \frac{1}{A}\sum_{i=1}^{n} A_i , \tag{2}$$

where V and A are a volume and an area of a reference part of a coating containing n microvoids; V_i and A_i are volumes and areas of individual microvoids, respectively.

Voids density ρ_a is defined as an average number of voids in a unit area and can be determined by micrographic analyses; then the *total number of microvoids* N_a in the area A is

$$N_a = \rho_a A . \tag{3}$$

The microstructure of ceramic coatings (see Figure 8) indicates that microvoids in coatings generally demonstrate irregular shapes but to simplify our analysis, the shape of microvoids is considered as an ellipse, which can be characterised using lengths of its major and minor axes. In general, these parameters – porosity, voids density/number, shape, size and orientation of microvoids – characterise the microstructure of ceramic coatings in this Section. The effective properties of coatings are significantly affected not only by the average values but also by a scatter in respective parameters, which are determined by means of image analyses of microstructure of coatings.

Image analyses for 10 cross-sectional micrographs (each 260 μm × 200 μm) of plasma sprayed alumina coatings allowed to determine areas and dimensions of 5252 microvoids (with the detection limit of 0.5 μm). The average porosity was 1.8% with the size of most microvoids being within the range 0 – 30 μm. Based on these data, all microvoids are divided into $n_{ss} = 59$ discrete types according to their shape and size obtained from the image analyses (see Figure 10).

The shapes of microvoids are divided into 16 bands using a *shape factor* S_f

$$S_f = 1 - \left(\frac{b}{a}\right)^2 , \tag{4}$$

where b and a are lengths of minor and major axes of the ellipsoid presenting the void, respectively (here all voids are assumed to have their major axis along

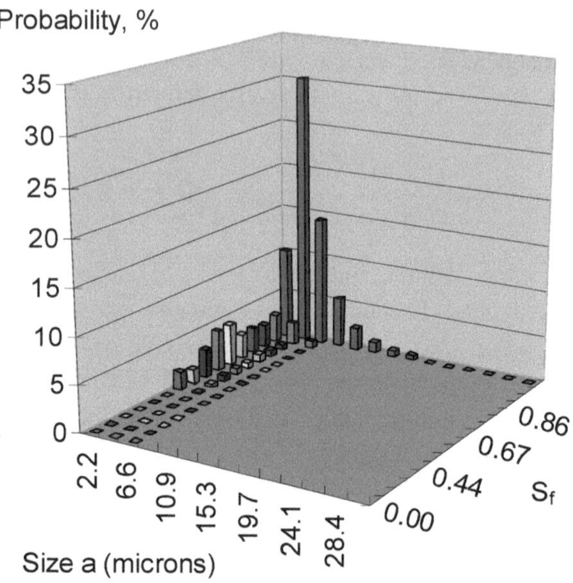

Figure 10. A shape-size-probability distribution of microvoids in plasma sprayed alumina coating with porosity 1.8%.

the transverse direction, i.e. in the deposition plane). The extreme values of the shape factor 0 and 1 correspond to circular voids and cracks, respectively. The size of voids, presented by a, is divided into 14 bands. Probability p_i for each type of voids in their distribution is determined as

$$p_i = \rho_i / \rho_a , \tag{5}$$

where ρ_i is a density of the i-th type of voids (defined as a number of voids in a unit area) and $\rho_a = \sum_{i=1}^{n_{ss}} \rho_i$ is the total void density of the coating.

In numerical simulations, the number N_i of the i-th type of microvoids in a representative element is determined by the Monte-Carlo algorithm. Then local effective mechanical parameters – the Young's moduli $\overline{E}_x$ and $\overline{E}_z$ in transverse directions and $\overline{E}_y$ in the spray direction as well as shear moduli and the Poisson's coefficient – of an element for materials with a transversely isotropic distribution of elliptical voids can be explicitly expressed in terms of the number of microvoids of various types, porosity p

$$p = \frac{1}{V}\frac{4\pi}{3}\sum_{k=1}^{n_{ss}} a_k^2 b_k N_k \tag{6}$$

and the crack density (hole density tensor $\boldsymbol{\beta}$ (Kachanov *et al.*, 1994 and Sevostianov *et al.*, 2004)

$$\boldsymbol{\beta} = \frac{1}{V}\sum_{k=1}^{n_{ss}} N_k\left(a_k^3\mathbf{nn} + b_k^3\mathbf{mm} + a_k^3\mathbf{ll}\right) \tag{7}$$

in the following way:

$$\overline{E}_1 = \overline{E}_2 = E_0\left[1 + \frac{32\left(1-\nu_0^2\right)}{3\left(2-\nu_0\right)}\frac{\beta_{11}}{1-p} + \frac{3\left(1-\nu_0\right)\left(9+5\nu_0\right)}{2\left(7-5\nu_0\right)}\frac{p}{1-p}\right]^{-1}, \tag{8}$$

$$\overline{E}_3 = E_0\left[1 + \frac{32\left(1-\nu_0^2\right)}{3\left(2-\nu_0\right)}\frac{\beta_{33}}{1-p} + \frac{3\left(1-\nu_0\right)\left(9+5\nu_0\right)}{2\left(7-5\nu_0\right)}\frac{p}{1-p}\right]^{-1}, \tag{9}$$

$$\overline{G}_{12} = G_0\left[1 + \frac{32\left(1-\nu_0\right)}{3\left(2-\nu_0\right)}\frac{\beta_{11}}{1-p} + \frac{15\left(3-\nu_0\right)}{2\left(7-5\nu_0\right)}\frac{p}{1-p}\right]^{-1}, \tag{10}$$

$$\overline{G}_{13} = \overline{G}_{23} = G_0\left[1 + \frac{16\left(1-\nu_0\right)}{3\left(2-\nu_0\right)}\frac{\beta_{11}+\beta_{33}}{1-p} + \frac{15\left(3-\nu_0\right)}{2\left(7-5\nu_0\right)}\frac{p}{1-p}\right]^{-1}, \tag{11}$$

$$\frac{\overline{\nu}_{13}}{\overline{E}_1} = \frac{\overline{\nu}_{31}}{\overline{E}_3} = \frac{\nu_0}{E_0}\left[1 + \frac{3\left(1-\nu_0\right)\left(1+5\nu_0\right)}{2\nu_0\left(7-5\nu_0\right)}\frac{p}{1-p}\right]. \tag{12}$$

In Eqs. (6)-(12) N_k is the number of the k-th type of voids in the total volume V; a_k and b_k are lengths of the major and minor axes of these voids; **n, m** and **l** are unit vectors of these axes; E_0, G_0 and ν_0 are the Young's modulus, shear modulus and Poisson's ratio of the undamaged material; $x \to 1$, $z \to 2$ and $y \to 3$.

A combination of the random allocation of voids to RVEs together with analytical relations for their local properties allows us to analyse the effect of

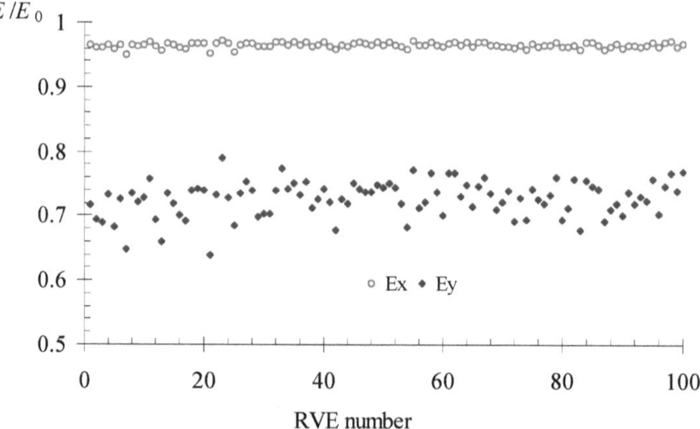

Figure 11. Normalized Young's moduli of plasma sprayed alumina coating with porosity 1.8%.

microstructure on the properties for RVEs of varying dimensions. The calculated local effective magnitudes of the Young's modulus for 100 random RVEs with dimensions 80 μm × 80 μm for alumina coating with average porosity 1.8% are shown in Figure11. The Poisson's ratio used in simulations is $\nu_0 = 0.27$. Obviously, the elastic properties in spray direction are much more sensitive to the microstructure – they demonstrate both the larger decline and the higher scatter.

The maximum scatter in sets of the effective properties diminishes with the increase in the dimensions of RVEs; this trend for the respective parameter for Young's moduli calculated as $\max\limits_i \left\{ \left| \overline{E}_i - \left\langle \overline{E} \right\rangle \right| \right\} / \left\langle \overline{E} \right\rangle$, $i = 1,...,100$, where $\left\langle \overline{E} \right\rangle$ is the averaged modulus, is obvious in Figure 12a. The calculations have demonstrated that the increase in the porosity level causes the increase in a relative scatter for the fixed dimensions of RVEs while the average mechanical parameters deteriorate (Figure 12b).

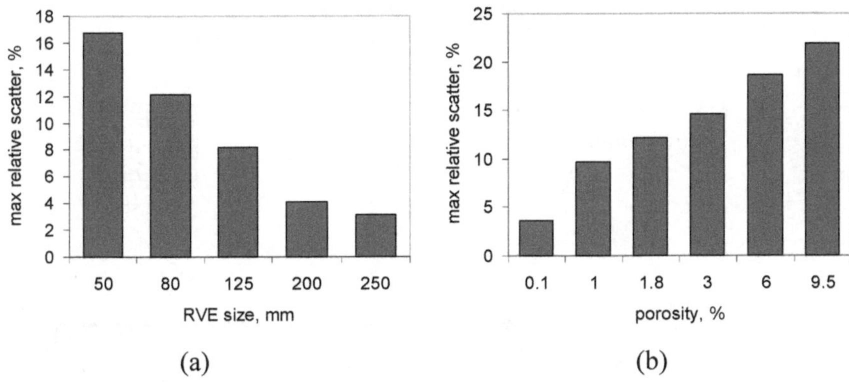

Figure 12. Change in scatter of the Young's modulus in spray direction: (a) with dimensions of RVE, porosity 18% and (b) with porosity, RVE size 80 μm.

3.2. Multi-scale strategy

To solve application problems for ceramic coatings, a two-scale (macro and meso) finite-element model was developed to analyse damage and fracture evolution. The macroscopic finite-element analysis is firstly conducted for a global region Ω^G (see Figure 13) exposed to external loading/environmental conditions that are implemented as respective boundary conditions in the finite-element simulations. At this level, macroscopic finite elements have material properties of a homogenised material, reflecting its averaged properties and global anisotropy.

A local region of interest Ω^L – for instance, the area near the focus of a laser beam in the problem of a laser-induced thermal shock (Zhao, 2005) – is then analysed at another level by means of a mesoscopic finite-element study. Material properties of each mesoscopic element are determined by its local heterogeneity defined by microstructural features such as microvoids, microcracks etc. based on the combination of the Monte-Carlo scheme, defining elements' microstructure, and analytical relations for mechanical properties discussed above.

The obtained results of the macroscopic simulations are used to define the boundary conditions for the boundary Γ^L of the local area Ω^L in mesoscopic analysis. These boundary conditions may be formulated in displacements or forces. For the former case, the applied boundary displacements are interpolated

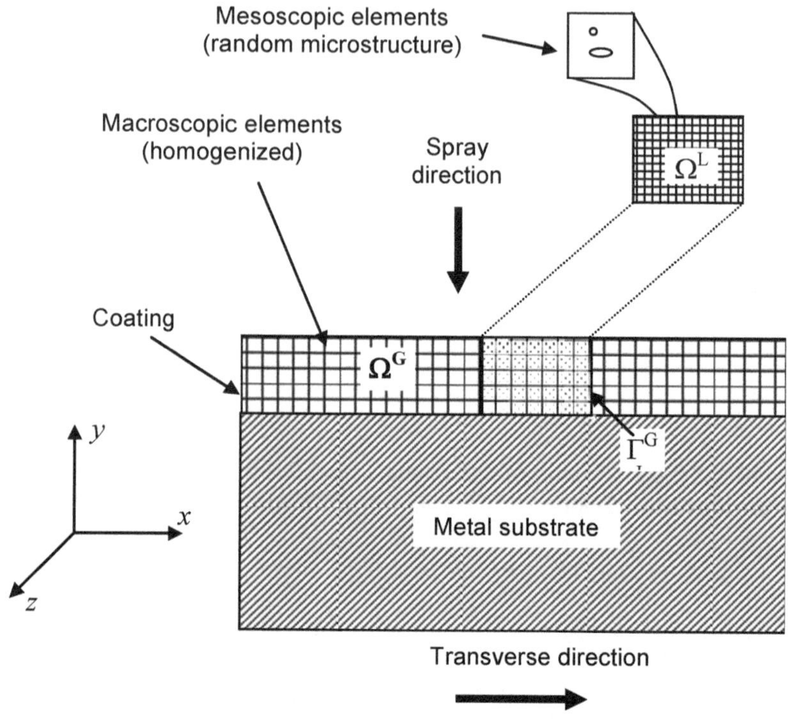

Figure 13. Two-scale finite-element model used to analyse damage and fracture in ceramic coatings.

from solutions for the macroscopic mesh. For the latter case, the internal forces or stresses obtained from the macroscopic calculation are converted to nodal forces on the mesoscopic mesh.

Obviously, the above description deals with the initial state of the material; a study of damage accumulation and fracture evolution presupposes introduction of additional mechanisms into consideration at the lower (mesoscopic) scale. In case of alumina coatings these elements include:

- Introduction of the initial level of damage linked to the random distribution of porosity in the coating;
- Damage evolution based on the original variant of continuum

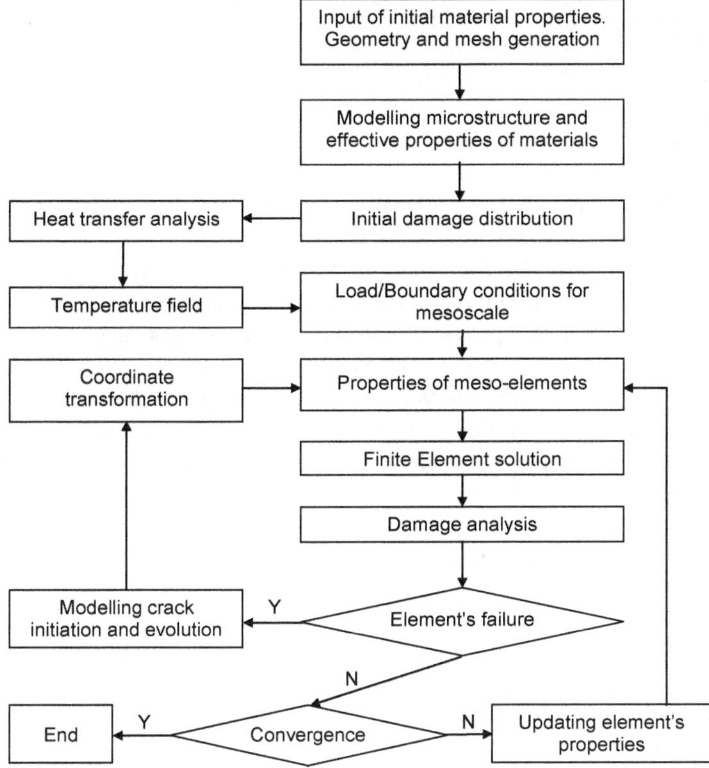

Figure 14. Flow chart of damage and fracture analysis for local region in porous alumina coating.

damage mechanics with a tensorial damage parameter (Najar and Silberschmidt, 1998 and Zhao, 2005) and respective deterioration of the local stiffnesses;

- Introduction of the local failure criterion that defines conditions for the crack initiation based on the critical magnitude of the major principal damage component;
- Account for damage-induced anisotropy by means of updating

stiffnesses of the elements with cracks (Zhao and Silberschmidt, 2005) – the local stiffness along the direction perpendicular to the crack (and coinciding with that of the major principal stress and damage components) is changed to a very small magnitude.

The flow chart in Figure 14 presents main elements of the computational multi-scale strategy used to simulate the response of a porous alumina ceramic deposited onto the metallic substrate to the laser-induced thermal shock. The solution at the macroscopic level was concentrated on the thermomechanical behaviour of the coating under the given temperature profile, measured in the experiment and using the effective thermomechanical properties. A transient thermomechanical formulation provided solutions used as boundary conditions for a mesoscopic scale at each time step. Then the damage evolution was calculated for this time step using the iteration procedure before the convergence criterion was met – as shown in the lower half of the flow diagram in Figure 14. Another criterion – linked with a possibility of the local failure – was applied to all the mesoscopic elements. If it was fulfilled for a given element, this element was considered as locally failed with a crack formed in it along the direction perpendicular to that of the major principal stress.

Obviously, a random microstructure results in significantly varying magnitudes of local damage accumulation rates. One of typical examples of damage distribution caused by such evolution is given in Figure 15. A somewhat chaotic image still demonstrates a proper capture of one of the observed features of damage in the coating – generation of a large crack near, and parallel to, the coating/substrate interface.

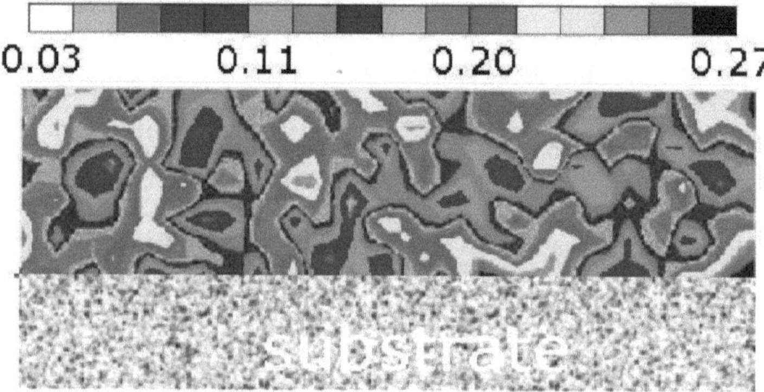

Figure 15. Damage distribution in local region of porous alumina coating.

4. Effect of Microstructure in Modelling Microelectronic Applications

Microelectronic devices are one of obvious examples where the microstructural features become an important factor, affecting their properties, performance and reliability. The major reason defining the prominence of microstructure is continuing miniaturisation in microelectronics with characteristic elements of packages, e.g. solder joints, having dimensions below 100 μm. In this case, a solder joint can consist of a single grain or several grains hence making the use of mechanical properties defined for bulk specimens at best questionable.

Another important feature of microelectronic packages is their multi-material compositions linked to the necessity to implement multiple functions (e.g. conductivity and isolation) that can not be achieved with the use of a single material. As a result, even a careful assembling aiming at stress-free packaging and protection from external loads can not fully avoid mechanical loading in microelectronics components due to inevitable thermal fluctuations and a mismatch in thermomechanical properties of constituents' materials.

For instance, in a flip-chip package, solder interconnections connect a substrate and a chip that have a pronounced mismatch of their coefficients of thermal expansion: that of an organic FR4 substrate is 18×10^{-6} K^{-1} while the one of Si used for chips is only 2.6×10^{-6} K^{-1}. Any temperature variations – either due to powering of the chip or caused by daily/seasonal changes – would result in (predominantly) shearing loading of solder joints as illustrated in Figure 16. Repetition of such changes is known as *thermal fatigue* that can eventually result in the failure of the package; this type of failure is one of the major ones in

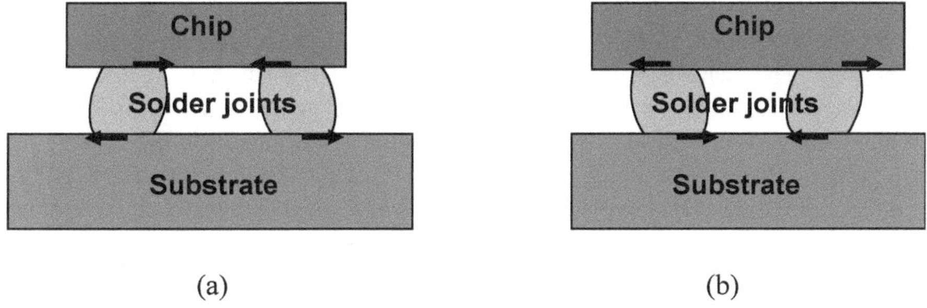

Figure 16. Shear strain in solder joints during purely thermal loading: (a) high temperature; (b) low temperature.

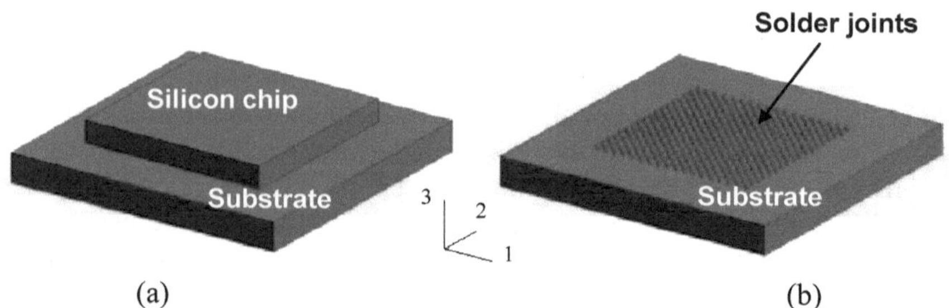

Figure 17. Geometry of flip chip package: (a) entire package; (b) package without silicon chip.

microelectronics.

Any analysis of reliability of such devices should be based on the data for fields of stresses and strains in order to define the places with large changes and to predict the life-in-service based on this data. A large difference in dimensions of microelectronic packages and their components, e.g. solder joints (see Figure 17), together with complex deformation processes at the microscopic scale in the latter prevents a direct introduction of an entire package together with the detailed microstructural descriptions into numerical simulations. Rather, a multi-scale simulation strategy can be used; two variants of such strategy are discussed in this Section following (Gong *et al.*, 2006a, 2007, and 2008 and Gong, 2007).

4.1 Effect of crystallinity

The first variant of multi-scale scheme deals with analysis of the effect of the crystalline structure of lead-free SnAgCu solder micro-joints on their creep behaviour. The main aim is to study the effect of transition from the use of mechanical properties obtained for a bulk specimen that has a microscopically isotropic behaviour to grain-defined ones. The latter are linked to the crystalline structure of β Sn – the major component of SnAgCu solder that has a body-centred tetragonal structure – resulting in locally anisotropic properties of its grains.

Two levels of finite-element simulations are implemented in this case. In the global model that describes the response to thermal cycling of the entire package using a fully coupled thermomechanical 2D formulation a relative displacement of the chip/joint interface with regard to the joint/substrate one is calculated. The obtained data is mapped onto a 3D sub-model of a finely-meshed solder joint as respective boundary conditions.

A crystal visco-plasticity model is used to model the behaviour of SnAgCu solder that was implemented by means of subroutine Umat in the commercial finite-element software package ABAQUS. In this model, the creep strain is defined as a sum of all component strains of the activated slip systems, which are controlled by the resolved shear stress on the corresponding slip plane.

The total strain is the sum of elastic, inelastic and thermal components; since the major effect of thermal loading in the assembly is accounted for at the level of a joint by introduction of boundary conditions from the global solution, the total strain rate $\boldsymbol{\varepsilon}_{\mathrm{total}}$ is defined here as

$$\boldsymbol{\varepsilon}_{\mathrm{total}} = \boldsymbol{\varepsilon}_{\mathrm{e}} + \boldsymbol{\varepsilon}_{\mathrm{c}}, \tag{13}$$

where $\boldsymbol{\varepsilon}_{\mathrm{e}}$ and $\boldsymbol{\varepsilon}_{\mathrm{c}}$ are elastic and creep strains, respectively. The anisotropic elastic behaviour is described by the Hooke's law

$$\boldsymbol{\sigma} = \mathbf{C} : \boldsymbol{\varepsilon}_{\mathrm{e}}, \tag{14}$$

where the fourth-order elastic stiffness tensor $\mathbf{C}$ has the following structure:

$$C_{ijkl} = G\left(\delta_{ik}\delta_{jl} + \delta_{il}\delta_{jk}\right) + \lambda\delta_{ij}\delta_{kl}\ . \tag{15}$$

Here G is the shear modulus, λ is the Lame's constant, δ_{ij} is the Kronecker symbol and $i, j, k, l =$ 1, 2, 3.

The evolution of creep strain is controlled by the contributions of shear creep rates by each slip system in SnAgCu (in total, 16 slip systems are considered in simulations):

$$\dot{\boldsymbol{\varepsilon}}_{\mathrm{c}} = \sum_{k=1}^{N} \dot{\gamma}_k \mathbf{s}_k\ . \tag{16}$$

where N is the number of slip systems, $\dot{\gamma}_k$ and $\mathbf{s}_k$ are the scalar shear strain rate the Schmid tensor of the k-th slip system, respectively. The latter can be expressed in the following form:

$$\mathbf{s}_k = \frac{1}{2}\left(\mathbf{m}_k \otimes \mathbf{n}_k + \mathbf{n}_k \otimes \mathbf{m}_k\right), \tag{17}$$

where $\mathbf{n}_k$ and $\mathbf{m}_k$ are unit vectors of the slip direction and the normal to the slip plane, respectively.

The shear strain rate is a function of the resolved stress on the slip system, temperature and hardening variables. Assuming the state of steady-state creep, the hyperbolic sine law is used to describe this behaviour:

$$\gamma_k = A_k \sinh(\tau_k)^n \exp\left(-\frac{Q_k}{RT}\right), \tag{18}$$

where A_k is a constant, n is a stress exponent, Q_k is the activation energy and τ_k is the resolved shear stress on the slip plane in the slip direction for the k-th system:

$$\tau_k = \mathbf{s}_k : \boldsymbol{\sigma} . \tag{19}$$

Several simplifying assumptions with regard to the grain-level mechanical behaviour are used in simulations: the effect of hardening is not considered; A_k and Q_k are considered to be the same for all systems; the effect of rotations is not accounted for as deformations are considered to be small.

The modelled area in the global 2D model is a middle section of the package normal to direction 3 in Figure 17; one half of this package is modelled due to its symmetry with respective boundary conditions at the symmetry axis. A thermal loading history starting by heating from the initial temperature 298 K to 373 K

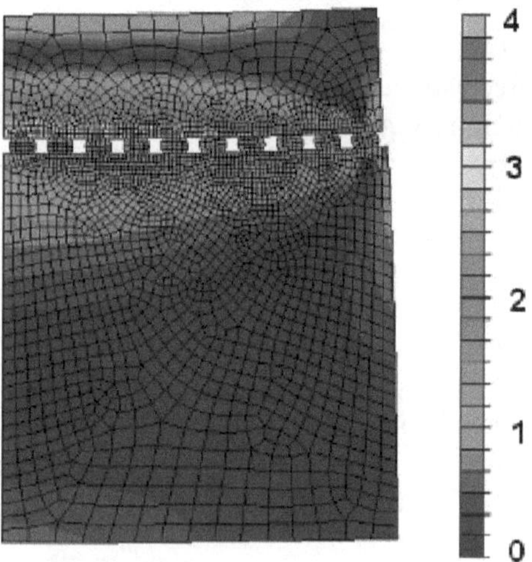

Figure 18. Distribution of equivalent stress (in MPa) and deformation of the global model after heating to 373 K.

and then cycling between the latter and 273 K is used in simulations; a spatially isothermal loading is used due to small dimensions of the package. The cooling and heating rates were constant with magnitude 10 K/min. A result of these package-level finite-element simulations for the moment with a maximum temperature after heating for 450 s is given in Figure 18.

The global analysis of deformational behaviour of the flip chip assembly under cyclic loading was used to define the micro-joint that undergoes the largest stresses and deformation in the package. Since it is most critical for reliability of the entire package, the detailed local analysis was implemented for it in order to study the effect of grain microstructure on its response to loading.

Simulations at the micro-scale were performed for four various types of the microstructure: one with properties of the bulk material and three other with various type of grain structure – single-crystal, bi-crystal and multi-crystal; the last two cases are shown in Figure 19. In this figure, colours correspond to different lattice orientations of the grains; to make analysis more transparent only three, mutually orthogonal, orientations of the body-centred tetragonal unit cell with $a = b > c$ were employed. In the case of the single-crystal joint, its three variants, each corresponding to one of these lattice orientations, are modelled.

The computational results for the first type of joints demonstrate that stresses at extreme temperatures concentrate at the chip/joint or joint/substrate interfaces with the highest stress concentration taking place at the edges of these interfaces

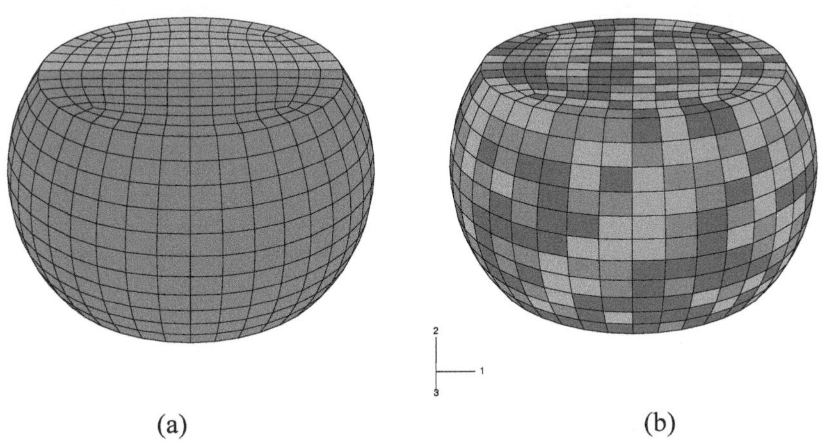

Figure 19. Crystalline structure in cases of bi-crystal (a) and multi-crystal (b) joints.

(Figure 20a).

This type of stress distribution can be linked to the shape of solder joints: it is

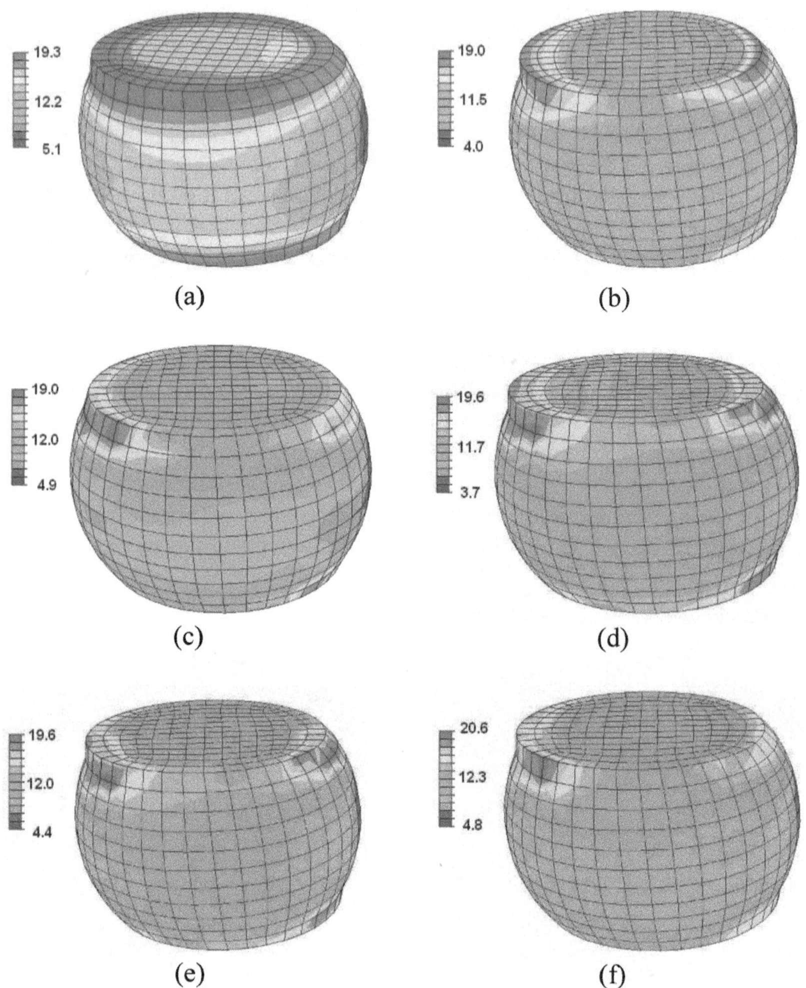

Figure 20. Distribution of equivalent stress (in MPa) and deformation of the global model after heating to 373 K in joints with various microstructure: (a) joint with isotropic properties of macroscopic material; (b)-(d) single-crystal joints with various lattice orientations, (e) bi-crystal joint; (f) multi-crystal joint.

not only the smallest cross section of joints but also the edges near the interfaces have an acute angle resulting in significant stress concentration (see also (Gong *et al.*, 2006b)).

This is also justified by experimental results demonstrating crack initiation exactly in these areas with subsequent crack propagation along the interface during thermal cycling. Hence, the stress and strain responses of these edges should be used to predict the fatigue life of joints.

In the models, accounting for crystallinity of joints, stresses still concentrate at the interfaces, with the highest stress levels occurring near the edges at these interfaces (Figure20b-f). This indicates that the shape of the joint is the major factor affecting its stress state. However, there are some obvious differences. For instance, stress gradients in the regions with stress concentration are significantly higher than in the case without accounting for grain orientation. This difference stems from the specificity of strain behaviour of the single crystal with inelastic strains occurring predominantly in certain slip systems and different directions of the creep strain rate and stresses. As a result, stresses in some directions are hard to release. Moreover, the higher the stresses the higher the residual stress that remains. This leads to additional stress concentration.

The mechanical behaviour of polycrystalline bulk specimens with a very large number of grains, resulting in quasi-isotropic properties, is equivalent to that of a single crystal with an infinite number of slip systems, which are randomly oriented in all possible directions. Hence, the high values of stresses are released by the preferably oriented slip system. This effective mechanism leads to lower stress gradients. There are no principal differences in stress distributions in single-grain joints with three different, mutually orthogonal, orientations of their lattices though magnitudes of stresses change (Figure 20b-d).

Transitions to a bi-crystal joint (Figure 20e) or a multi-crystal one (Figure 20f) does not affect considerably the character of stress distribution (in both cases the effect of grain boundary is neglected). Hence, at the micro-scale, the structure of the electronic package, e.g. geometry of its joints, is still a major factor that determines its reliability. Still, numerical simulations demonstrated that the type of crystallinity of solder joints affects both magnitudes and the character of distribution of various stress components in them and hence should be accounted in the reliability study of microelectronic packages.

4.2 Effect of deformational mechanisms

Development of another variant of the multi-scale model was linked to two main objectives:

- To introduce a 3D variant of the global model of the flip-chip package;
- To introduce additional deformational mechanisms in the constitutive description used in simulations.

To implement the first objective, a quarter of the assembly was discretised into 3D finite elements (Figure 21) to simplify the model that have two planes of symmetry; respective boundary conditions were introduced for those planes. The model geometry for the package is based on dummy components provided by a supplier. The silicon chip has the dimensions 5.08 mm × 5.08 mm with thickness 0.625 mm. An area array of joints consists of solder bumps with diameter of 136 μm and height 100 μm; the spacing is 254 μm. The four-layer FR4 printed circuit board has thickness of 0.8 mm.

One joint in the centre of the model (marked with an arrow in Figure 21b) is used for the local analysis and, therefore, has a finer mesh (Figure 22) than other bumps in the array. The under-bumps metallization between the silicon chip and solder joints and metal finishes between solder joints and the substrate are neglected in the model, and three components are ideally bonded along the interface (each with diameter 110 μm).

The materials description is extended to accommodate another creep mechanism linked to void diffusion and the effect of thermal deformations at the microscopic scale (Gong *et al.*, 2008 and Gong, 2007). Hence, the total strain rate in this case can be presented as

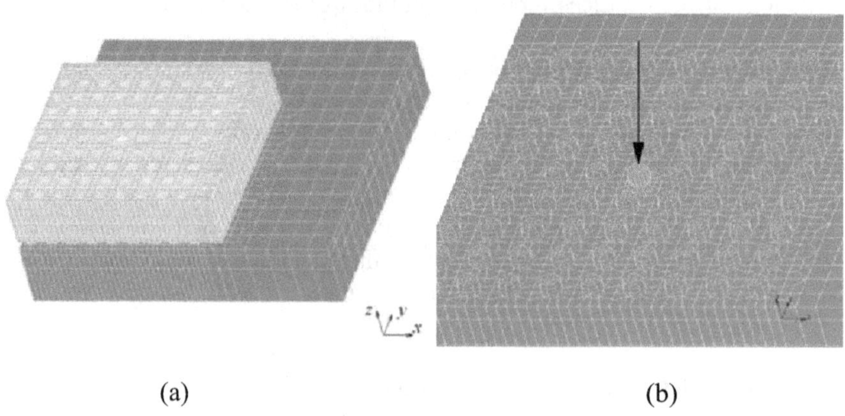

(a) (b)

Figure 21. 3D finite-element mesh for flip chip package: (a) quarter of entire package; (b) substrate with joints.

$$\dot{\boldsymbol{\varepsilon}}_{\text{total}} = \dot{\boldsymbol{\varepsilon}}_{\text{e}} + \dot{\boldsymbol{\varepsilon}}_{\text{c}} + \dot{\boldsymbol{\varepsilon}}_{\text{th}}, \tag{20}$$

where the old subscripts are the same as in Eq. (13); the subscript 'th' denote thermal components. To accommodate the additional deformational mechanism, the inelastic strain term now has the following form:

$$\boldsymbol{\varepsilon}_{\text{c}} = \boldsymbol{\varepsilon}_{\text{md}} + \boldsymbol{\varepsilon}_{\text{vd}}. \tag{21}$$

Here the term describing dislocation movement in slip planes – $\boldsymbol{\varepsilon}_{\text{md}}$ – has the same form as before (see Eq. (16)); but another expression was used to describe the shear strain rate for the steady-state creep – a power law:

$$\dot{\gamma}_k = A_k \left(\tau_k\right)^{n_1} \exp\left(-\frac{Q_k}{RT}\right), \tag{22}$$

where n_1 is the stress exponent.

The strain due to the vacancy diffusion – $\boldsymbol{\varepsilon}_{\text{vd}}$ – is introduced in terms of the isotropic intragranular stress-controlled diffusion; the respective creep rate can be presented with another power law:

$$\dot{\varepsilon}_{\text{vd}}^{ij} = \frac{3s^{ij}}{2\sigma_{\text{eq}}} A_{\text{vd}} \left(\sigma_{\text{eq}}\right)^{n_2} \exp\left(-\frac{Q_{\text{vd}}}{RT}\right). \tag{23}$$

Here, s^{ij} is a deviatoric stress component; σ_{eq} is the equivalent stress; n_2 is the

Figure 22. 3D finite-element mesh of the joint used for local analysis and its surroundings.

stress exponent; A_{vd} and Q_{vd} are the material constant and the activation energy for VD, respectively.

The last term in Eq. (20) – the thermal strain rate – should, in a general case, reflect anisotropy of micro-joints due to their crystallinity. So, instead of a single scalar a second-order tensor of coefficients of thermal expansion $\boldsymbol{\alpha}$ is used:

$$\dot{\boldsymbol{\varepsilon}}_{th} = \boldsymbol{\alpha}\dot{T}\,. \tag{24}$$

A special procedure was developed for this new description (for details see Gong *et al.* (2008) and Gong (2007)); a user-defined subroutine Umat was programmed to implement the suggested algorithm in finite-element simulations. The same loading history as in the simulations with the previous model (Section 4.1) was used in the analysis.

At the first stage of numerical simulations, the entire package (i.e. its quarter) is modelled; all solder joints are presented with mechanical properties of a bulk solder specimen. A distribution of the equivalent stress in the package is presented in Figure 23 at temperature 318 K, after two minutes of heating from the initial temperature.

The figure clearly demonstrates (though exaggerated by the used scale factor for deformation) the warpage of the substrate. This behaviour is linked to the difference in the levels of CTEs of the silicon chip and substrate. Due to stiffness

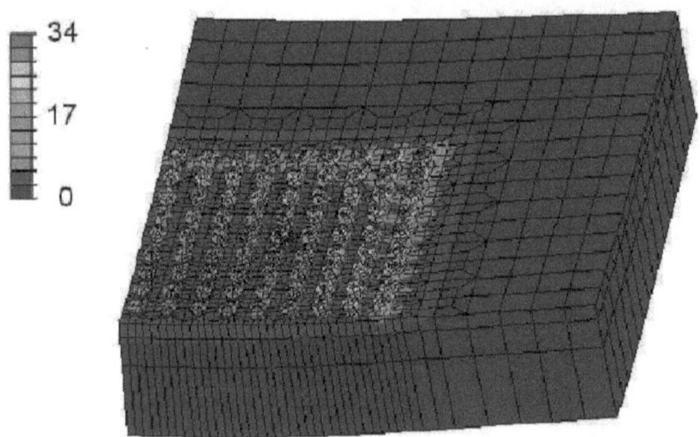

Figure 23. Equivalent stresses in package (with removed silicon chip) after temperature increases to 318 K in the first cycle. Scale factor for deformation is 300.

of the chip (its Young's modulus is 131 GPa), when the temperature increases the deformation is transferred through solder joints and absorbed by the soft substrate (part of it is also absorbed by joints). The constraint on expansion of the substrate's top surface is comparatively large, leading to the upward warpage.

The deformation of the chip is similar to that of the substrate, but with a significantly lower amplitude. Solder joints between the substrate and the silicon chip experience a shearing deformation, which increases with the distance form the chip's centre. Therefore, under thermal cycling, interfaces of peripheral joints are one of the most critical positions in the package.

At the local level, the solder bump with refined meshing is treated as a single-crystal joint. In the crystal model, all the strain components, including elastic, creep and thermal expansion, are accounted for; 16 activated slip systems (the same as in the previous model) are considered. A distribution of the equivalent stress in the single-crystal joint after the temperature increase to 318 K from its initial level is shown in Figure 24a. Obviously, there are two areas of stress concentration at the chip/joint interface on opposite edges along the direction of the maximum shearing deformation.

The finite-element simulation tools allows to study single deformational mechanisms and their effect on response of solder joints to thermal cycling that in real-life experiments would be rather cumbersome or even impossible. For instance, in order to investigate the contribution of thermal expansion, the

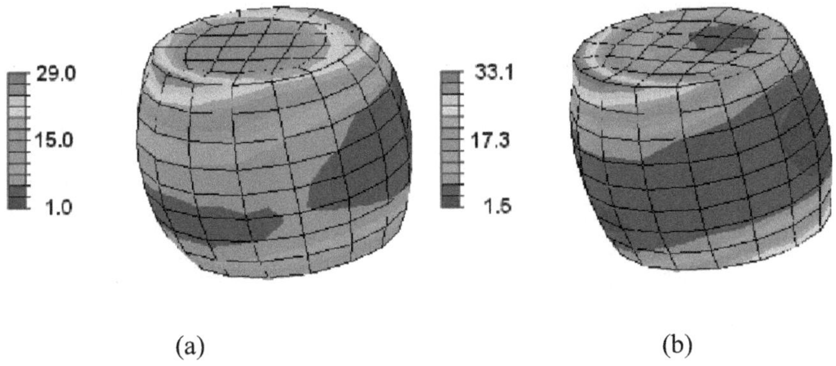

Figure 24. Calculated distributions of equivalent stresses in single-crystal joint at 318 K: (a) with account for thermal expansion; (b) with thermal expansion de-activated.

thermal component of the constitutive model is deactivated in simulations of the single-crystal joint. Other joints, as well as the silicon chip and the substrate, retain this component.

The resulting distribution of equivalent stress for conditions of Figure 24a (excluding the effect of thermal expansion) are presented in Figure 24b. The effect of de-activation is apparent – instead of two stress concentration areas at the chip/joint interface there is only one, with another appearing at the joint/substrate interface.

It is also possible to estimate the inputs of both non-elastic deformational mechanisms. For this purpose, the equivalent strain

$$\varepsilon_{eq} = \sqrt{\frac{2}{3}\boldsymbol{\varepsilon} : \boldsymbol{\varepsilon}} \tag{25}$$

is calculated for components linked to movement of dislocations and vacancy diffusion; the obtained distributions for a single-crystal joint at 318 K after first two minutes of heating are presented in Figure 25. It can be seen that deformations due to VD are considerably smaller since higher temperatures are needed for accelerated deformations due to this mechanism. The simulations show that even at the highest temperature of the cycle – 373 K – this mechanism has little effect.

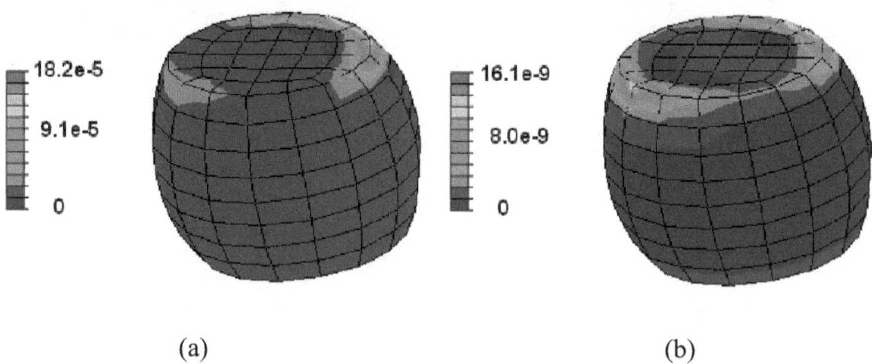

Figure 25. Calculated distributions of equivalent strains in single-crystal joint at 318 K due to: (a) movement of dislocations; (b) vacancy diffusion.

Acknowledgement

The author would like to acknowledge contributions of the members – former and current – of the Mechanics of Advanced Materials Research Group at Loughborough University.

Bibliography

P. W. R. Beaumont, R. A. Dimant and H.R. Shercliff. Failure processes in composite materials: getting physical, *Journal of Materials Science.* 41:6526–6546, 2006.

J.-M. Berthelot. Transverse cracking and delamination in cross-ply glass-fiber and carbon-fiber reinforced plastic laminates; static and fatigue loading. *Applied Mechanics Reviews*, 56:111–147, 2003.

R. J. Damani and A. Wanner. Microstructure and elastic properties of plasma-sprayed alumina. *Journal of Materials Science*, 35:4307-4318, 2000.

J. Gong. *Microstructural Features and Mechanical Behaviour of Lead Free Solders for Microelectronic Packaging.* PhD Thesis, Loughborough University, 2007.

J. Gong, C. Liu, P. P. Conway and V. V. Silberschmidt. Analysis of stress distribution in SnAgCu solder joint. *Applied Mechanics and Materials,* 5-6:359-366, 2006a.

J. Gong, C. Liu, P. P. Conway and V. V. Silberschmidt. Modelling of Ag3Sn coarsening and its effect on creep of Sn–Ag eutectics. *Materials Science and Engineering A*, 427:60–68, 2006.

J. Gong, C. Liu, P. P. Conway and V. V. Silberschmidt. Micromechanical modelling of SnAgCu solder joint under cyclic loading: Effect of grain orientation. *Computational Materials Science*, 39:187–197, 2007.

J. Gong, C. Liu, P .P. Conway and V. V. Silberschmidt. Mesomechanical modelling of SnAgCu solder joints in flip chip. *Computational Materials Science*, 43:199–211, 2008.

C. T. Herakovich. *Mechanics of Fibrous Composites*. John Wiley & Sons, 1988.

M. Kachanov, I. Tsukrov and B. Shafiro. Effective moduli of solids with cavities of various shapes. *Applied Mechanics Reviews*, 47:151-174, 1994.

Z. R. Khokhar, I. A. Ashcroft and V. V. Silberschmidt. Matrix cracking and delamination in cross-ply laminates in tensile fatigue. In *Proc. 13th European Conference on Composite Materials*, 2–5 June 2008, Stockholm, Sweden (electronic publication).

Y. W. Kwon, D. H. Allen and R. Talreja (eds.). *Multiscale Modeling and Simulation of Composite Materials and Structures*. Springer, 2008.

D. L. McDowell. Viscoplasticity of heterogeneous metallic materials. *Materials Science and Engineering R*. 62:67–123, 2008.

J. Najar and V. V. Silberschmidt. Continuum damage and failure evolution in inhomogeneous ceramic rods. *Archive of Mechanics*, 50:21-40, 1998.

K. L. Reifsnider and S. W. Case. *Damage tolerance and durability of material systems*. Wiley-Interscience, 2002.

J. Y. Rho, L. Kuhn-Spearing and P. Zioupos. Mechanical properties and the hierarchical structure of bone. *Medical Engineering & Physics*. 20:92-102, 1998.

I. Sevostianov, M. Kachanov, J. Ruud, P. Lorraine and M. Dubois. Quantitative characterization of microstructures of plasma-sprayed coatings and their conductive and elastic properties. *Materials Science and Engineering A*, 386:164-174, 2004.

V. V. Silberschmidt. Scaling and multifractal character of matrix cracking in carbon fibre-reinforced cross-ply laminates, *Mechanics of Composite Materials and Structures*, 2:243-255, 1995.

V. V. Silberschmidt Model of matrix cracking in carbon fiber-reinforced cross-ply laminates. *Mechanics of Composite Materials and Structures*, 4:23-37, 1997.

V. V. Silberschmidt. Multifractal characteristics of matrix cracking in laminates under T-fatigue. *Computational Materials Science*, 13:154-159, 1998.

V. V. Silberschmidt. Multi-scale modelling of cracking in cross-ply laminates. In C. Soutis and P. W. R. Beaumont, editors, *Multi-scale Modelling of Composite Material Systems: The Art of Predictive Damage Modelling.. Cambridge*. Woodhead Publishing, 2005a.

V. V. Silberschmidt. Matrix cracking in cross-ply laminates: effect of randomness. *Composites Part A - Applied Science and Manufacturing*, 36:129–135, 2005b.

V. V. Silberschmidt. Effect of micro-randomness on macroscopic properties and fracture of laminates, *Journal of Materials Science*, 40:6768-6776, 2006.

V. V. Silberschmidt. Account for random microstructure in multiscale models. In Y. W. Kwon, D. H. Allen and R. Talreja, editors, *Multiscale Modeling and Simulation of Composite Materials and Structures*. Springer, 2008.

C. Soutis and P. W. R. Beaumont (eds.). *Multi-scale Modelling of Composite Material Systems: The Art of Predictive Damage Modelling*. Woodhead Publishing, 2005.

A. Wanner and E. H. Lutz. Elastic anisotropy of plasma-sprayed, free-standing ceramics. *Journal of the American Ceramic Society*, 81:2706-2708, 1998.

J. Zhao and V. V. Silberschmidt. Microstructure-based damage and fracture modelling of alumina. *Computational Materials Science*, 32:620-628, 2005.

J. Zhao and V. V. Silberschmidt. Micromechanical analysis of effective properties of plasma sprayed alumina coatings. *Materials Science and Engineering A*, 417:287–293, 2006.

J. Zhao, *Modelling Damage and Fracture Evolution in Plasma Sprayed Ceramic Coatings: Effect of Microstructure*. PhD Thesis, Loughborough University, 2005.

Zeitfracht Medien GmbH
Ferdinand-Jühlke-Straße 7
99095 Erfurt, Deutschland
produktsicherheit@kolibri360.de